AF506007

Leprosy in Colonial South India

Leprosy in Colonial South India

Medicine and Confinement

Jane Buckingham
Lecturer in South Asian History
University of Canterbury
Christchurch
New Zealand

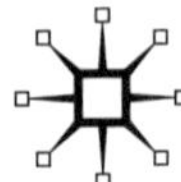

First published 2002 by
PALGRAVE
Houndmills, Basingstoke, Hampshire RG21 6XS and
175 Fifth Avenue, New York, N. Y. 10010
Companies and representatives throughout the world

PALGRAVE is the new global academic imprint of
St. Martin's Press LLC Scholarly and Reference Division and
Palgrave Publishers Ltd (formerly Macmillan Press Ltd).

ISBN 0–333–92622–6

This book is printed on paper suitable for recycling and
made from fully managed and sustained forest sources.

A catalogue record for this book is available
from the British Library.

Library of Congress Cataloging-in-Publication Data
Buckingham, Jane.
    Leprosy in colonial south India : medicine and confinement /
    Jane Buckingham.
        p. cm.
    Includes bibliographical references and index.
    ISBN 0–333–92622–6
    1. Leprosy—India—History. 2. Leprosy—Government policy–
    –India—History. 3. Leprosy—Social aspects—India—History.
    4. Imperialism—Health aspects—India—History. 5. Public health–
    –Political aspects—India—History. 6. Medicine—India—History.
    I. Title.

    RA644.L3 B83 2001
    614.5'46'0954—dc21
                                                    2001032726

10   9   8   7   6   5   4   3   2   1
11   10   09   08   07   06   05   04   03   02

Printed and bound in Great Britain by
Antony Rowe Ltd, Chippenham, Wiltshire

For my parents

Contents

Acknowledgements

This book began life as a doctoral thesis and during its research, writing and revision I have been encouraged and assisted by countless people at a professional and personal level. These acknowledgements are but a poor effort to express something of my deep gratitude and affection for those who made it possible.

I owe many intellectual and personal debts to Professor Ambirajan, David Arnold, Padma Anagol-McGinn, Richard Barz, Chris Bayley, Dipesh Chakrabarty, Indira Choudhury Sengupta, Philip Constable, Lesley Hall, Phillipa Levine, Jim Masselos, S.N. Mukherjee, V.R. Muraleedharn, Geoff Oddie, Deryck Schreuder, Michael Shortland, Judith Snodgrass, Gaurie Viswanathan and Anand Yang. My grateful thanks to Helen Frazer, Gary Hausman, Eugene Irschick and Dr Kandiah for guidance in Tamil language study and sources and my special thanks to Dr A.R. Venkatachalapathy for his friendship and his superb Tamil translations.

Australian libraries were my first research source. I must thank the staff of the National Library of Australia, the New South Wales State Library, the State Library of Victoria, and the Libraries of the Australian National University, New South Wales University and the University of Sydney. At the University of Sydney, inter-library loans staff and staff of the Medical and Public Health Libraries were particularly helpful. Staff of the Brownless Medical Library at the University of Melbourne were also of assistance. I am especially grateful to Brenda Heagney, the Librarian of the Royal Australasian College of Physicians, Sydney, Library for her expertise and extraordinary efforts in tracking down medical journals in Australia.

I am grateful for the patient assistance of the staff at the Oriental and India Office Collections, the British Library, the Libraries of the School of Oriental and African Studies and the School of Hygiene and Tropical Medicine, University of London, the Royal College of Surgeons of England Library and Cambridge University Library. Special thanks are due to Mary Douglas of the School of Hygiene and Tropical Medicine for her suggestions and perseverance in pursuing elusive materials. I owe a particular personal and academic debt to the library and archive staff and the students of the Wellcome Institute for the History of Medicine, London.

I am deeply grateful to Valerie Crust and the director and staff of The Leprosy Mission International's headquarters in London for opening their archives and library to me and for their encouragement and inspiration. I owe a special debt to Bill Edgar, formerly director of the Mission, who first welcomed me to the London headquarters in 1990, and whose death in 1992 was a sad loss. My grateful thanks go also to Dr Walter Cornelius who, as director of the Leprosy Mission, India, in 1990 eased my shock at seeing advanced leprosy and suggested I contact the Mission in London. I am grateful also to the staff of ILEP, International Federation of Anti-Leprosy Associations, and in particular to Sarah Lacey for her assistance and friendship in India and London. Special thanks to my friends Rachel and Isin and Santo and Virginia for support and shelter in London and to Santo for some troublesome sleuthing.

I am indebted to the staff of the National Archive of India and the Nehru Memorial Library in New Delhi, and my many friends at the Tamil Nadu State Archive and Library for their generous assistance over the years and to Givana Sekari, Librarian at the Central Research Institute (Siddha), Madras. Thanks are owed to the staff of the Shieffelin Leprosy Research Centre, Karigiri, and the Gremalters Referral Hospital and Leprosy Centre, Madras. My thanks also to Paul and Margaret Brand, Ernest P. Fritschi and the late T.N. Jagadisan for giving me insight into the grim realities of leprosy.

My grateful thanks also to those who were part of my daily life in Madras – Sally, Susie, Catherine, Sarah, Susheela, Kanimolhi, Dr T.K. Shankar, Andrew Lightman and Mike McFarlane, to my many friends at the YWCA Madras, and to the Augustine and Pius families. My thanks to Arjun and Amrita Puri, Gaurab and Swapna Banerjee, in Delhi, to Hans and to my friends in Hyderabad, particularly, Karla Karthikayan, Tom, Fritz and Jay Raj for their hospitality and care in sickness and health.

I am indebted to the Australian government for providing a substantial Australian Award for Research in Asia and the British Council for giving me funding under the Academic Links and Interchange Scheme. Grateful thanks are due to the Departments of History, University of Sydney, and of Philosophy and Religious Studies, and History, University of Canterbury, for providing facilities and support in the preparation of this manuscript. Special thanks to Charles Campbell and Sarah Coleman for helping to finalize the manuscript and index for publication. I am particularly indebted to Janey Fisher for the index and to both Janey and Palgrave for their considerate assistance throughout the publication process.

My most recent debt of gratitude is to my friends at Canterbury and colleagues in the History Department for their encouragement and friendship, particularly to S.A.M. Adshead for, among many things, reading the manuscript.

My love and thanks to my family and to Bert Balding for his gift, Simon Combe, Susan Conley, Phill and the Harts, Carolyn Hayes, Adrian Lyons, Abay Puri and Kajri Jain, Térèse Taylor, Elizabeth Weiss and Coralie Younger.

This book was written for leprosy sufferers and for those involved in their care, to give some sense of their history and to show how far they have come. I only wish it were a better offering.

Introduction

Leprosy is a bacterial infection caused by *Mycobacterium leprae*, an organism closely related to the tuberculosis bacterium. Unlike tuberculosis, leprosy is difficult to contract and is rarely fatal. The exact mechanisms for the communication of leprosy are still unknown, though it is believed that while bacteria can be discharged through leprous ulcers, the principal mode of transmission is by coughing and sneezing. Contrary to popular belief, leprosy is neither sexually transmitted nor is it inherited. Although they can infiltrate every part of the body, leprosy bacteria concentrate in the skin and peripheral nerves, resulting in the dramatic dermatological effects of the disease. The primary symptoms of leprosy – paralysis, inflammation and loss of feeling – are the result of nerve damage caused by the body's immune response to bacterial invasion. Though rare in Europe, leprosy is still prevalent in India and Africa, and there are currently over one million cases in India. Leprosy is now curable with multi-drug therapy, which involves the use of various combinations of anti-bacterial drugs and anti-inflammatory drugs which can reduce nerve damage and thus prevent the paralysis and anaesthesia which usually follow.

In the West leprosy is regarded as an ancient disease, familiar from Leviticus and the Christian tradition, but rarely seen. Even so, the disease has a powerful symbolic resonance and the term 'leper' is still commonly used as a synonym for a person feared, hated and driven to the margins of society. I have used the term 'leprosy sufferer' wherever possible in an effort to avoid the stigmatizing term 'leper'.

Stigmatism, while provoking horror, also provokes curiosity. However, this volume is driven more by curiosity about the dynamics of colonial power, specifically the contribution of medicine and law to the configuration of colonial authority. It concerns particularly the relationship between indigenous and British medical and legal systems, and the impact of such systems on the leprosy sufferer in south India. South India, more particularly the Madras presidency encompassing much of contemporary Kerala, Tamil Nadu and parts of Andhra Pradesh, is the focus of this study for two reasons. During the British colonial period, of the three British presidencies the Madras presidency consistently had the lowest proportion of leprosy sufferers. However, the Madras presidency provided more government-supported facilities for the care and treatment of leprosy than any other province of India including the Bengal and Bombay presidencies. While considering questions of policy and practice, the mutual influence of British and indigenous medicine, the international and Indian development of colonial science, the professionalization of medical and sanitary care and the shift to a robust legislative response to leprosy sufferers, the core questions of this book relate to the identity and life experience of the leprosy sufferer. Who suffered leprosy in colonial south India? What were the indigenous and the British legal and medical responses? How did leprosy sufferers live in relation to these responses? Although Foucault ascribes to the 'leper' a status of social exclusion and sub-ordination to external authority,[1] leprosy sufferers in colonial south India, despite their small number, had at least the power of resistance and, to a lesser extent, negotiation.

Focusing on colonial Madras presidency within the broader Indian and international context, this investigation of leprosy breaks fresh ground in a field dominated by Eurocentric enquiry into the bibli-cal tradition of leprosy, the relationship of leprosy to medieval Christianity and its inspiration for nineteenth-century missionary activity. Work by Iliffe and Vaughan on leprosy in Africa has con-centrated on the disease and its sufferers as part of Christian history. For Iliffe, the history of African leprosy is a triumphal story of medical and missionary hope and faith, marked by the 'dedication of those relieving it, and the courage of its victims'.[2] For Vaughan, however, it represents 'the projection on to Africa of a powerful Christian disease symbolism and the attempt to engineer socially a "leper identity" in the particular circumstances of colonialism'.[3]

Until the turn of the century, leprosy in India was regarded as a medical and government matter rather than a missionary responsi-bility. This book is concerned primarily with the period prior to substantial missionary involvement in leprosy in India, before nineteenth-century missionaries embraced leprosy sufferers as 'biblical lepers' and before the idea of the physically corrupted leprous body became, in the colonial world, a metaphor for the sinner's corrupted soul.[4]

Here leprosy is investigated from the beginnings of British insti-tutional care of leprosy sufferers in the Madras presidency in the early 1800s to the passing of the Government of India's Lepers Act in 1898. It is a period of significant transition in India's medical, legal and political life. In the aftershock of the 1857 Indian Mutiny, British power in India was officially transferred from the East India Company to the British Crown. Crown rule was characterized by increasing monitoring of the sanitary, medical and social conditions of the Indian populace. Such information-gathering was primarily intended to assist in protecting the health and well-being of the British army in India and thus also that of the British civilian population. Gradually indigenous health too became incorporated into a British imperial agenda shaped in part by a sense of responsibility for the welfare of India and its peoples. Among other measures, protection for both British and Indian health took the form of cautious but persistent efforts to legislate controls on epidemics and endemic forms of disease such as leprosy. Coupled with these changes, from the mid-nineteenth century there was a fundamental shift in British medicine as practised in India and at home, from a humoral to a scientific understanding of disease,[5] accompanied by the professionalization of medical practice and the bureaucratization of medical research.

In the 1891 Census of India, only Mysore and the Punjab returned a lower proportion of leprosy sufferers per 10 000 of the population than the Madras Presidency.[6] Even so, by the mid-nineteenth century, British leprosy care in Madras, in terms of the number of leprosy asylums available, the number of sufferers in care and the amount spent on their treatment and shelter, was more advanced than in any other province of India. By the last decade of the nineteenth century, only Bombay had caught up, with the rest of India lagging far behind.[7] In the city of Madras, part of the Tamil-speaking region of the Madras presidency, the increasing British government attention to leprosy care can be traced from its early-nineteenth century origins. The Madras presidency thus provides the fullest record of British colonial government-supported activity directed towards leprosy care.

Further, in the south the indigenous Ayurvedic medicine of the Sanskrit tradition, dominant in north India, is complemented and often supplanted by the vigorous southern Tamil tradition of Siddha medicine. The focus of this study on the Madras presidency thus allows an analysis which juxtaposes a strong traditional regional indigenous response to leprosy with negotiation within and between levels of local and central government authority in the effort to provide leprosy care.

The Siddha medical tradition has been little considered in historical writing on indigenous and colonial Indian medicine. This neglect is part of a preference in Indian historiography for studies of the sub-continent north of Hyderabad. Greater attention has been paid to developing histories of the northern regions held under Mughal rule prior to the British presence and the building of British colonial state formations in the Bengal and Bombay presidencies on the shoulders of the Mughal empire. The Madras presidency was largely untouched by Mughal influence and, as British power gained ascendency, remained the furthest presidency from the seat of colonial authority in Calcutta. As such, the Madras presidency offers a relatively little-researched field for the investigation of colonial/indigenous relationships articulated through mutual engagement with the medical, legal and social implications of a severe endemic disease.

The history of medicine in colonial societies, particularly colonial south Asia, is attracting increasing interest. However, work by Arnold, Catanach and others has focused primarily upon epidemic diseases such as smallpox, cholera and plague which, because of their impact on European civilians and the army, were perceived to be a significant threat to British military and symbolic power and attracted a vigorous government response. As a slow, degenerative disease, believed by the British to afflict Indians and Eurasians rather than their own people, leprosy had a more subtle engagement with colonial power. As such, the history of leprosy offers a unique perspective on the nature and development of colonial medical intervention. It allows a reconsideration of models of colonial medical relationships developed from the shattering effects of epidemic disease and can contribute to our emerging understanding of public health.

Other histories of disease, medicine and public health in India have emphasized the development of colonial state medicine and its relationship to indigenous society under colonial rule, particularly the degree of state intervention in the health of the Indian populace. Arnold and Harrison, while not ascribing to medicine the same status as a 'tool of empire' conferred by Headrick, counter Ramasubban's assertions of colonial negligence in the face of Indian health needs.[8] Rather, they suggest European medicine did intervene to assist Indians towards better health, expanding from its 'enclavist' focus on the needs of Europeans to a wider public health system including, albeit inadequately, the Indian populace.[9] The history of leprosy confirms British attention to Indian health, particularly at the lowest levels of society. Members of the Indian

Medical Service cared for the marginalized and infirm poor including many leprosy sufferers of all races in hospitals, jails, lunatic asylums, and civil hospitals and dispensaries.

Given the paucity of historical writing on leprosy in India (though there is a substantial body of anthropological and sociological material), this book is necessarily chronological and descriptive while engaging with aspects of social, medical and legal history. A major theme of this work is the expression of colonial power through law, specifically through the use of police power to confine vagrant and poor leprosy sufferers. The consideration of the confinement of leprosy sufferers in south India is influenced by recent interest in the history of prisons and institutional confinement, as seen in David Garland's *Punishment and Modern Society* and Ignatieff's *A Just Measure of Pain*, which both build upon and critique Foucault's seminal works, *Madness and Civilisation* and *Discipline and Punish*. Waltraud Ernst's *Mad Tales of the Raj*, on confinement of European 'insanes', is the only major study to date of medical confinement of a marginal population in India. Confinement of women practising prostitution was more central to the maintenance of the British Empire's military power, and has attracted more historical interest than has the confinement of either the insane or leprosy sufferers.

The nature of colonial power is not, however, to be discovered only in policing. The relationship of British and indigenous medicine in India, particularly the extent of British medical domination, has recently emerged as an important theme in the literature on colonial medicine. Bala argued in *Imperialism and Medicine in Bengal*, that Western medicine dominated indigenous medicine there by the turn of the nineteenth-century. European medicine did have a considerable impact in south India. However, there is strong evidence of the resilience of indigenous leprosy treatments, both in the south and internationally, into the twentieth-century. Regardless of the rhetoric of British medical superiority, effectiveness was the criterion for the success of indigenous remedies over British remedies for leprosy. Further, just as the co-operation of leprosy sufferers was necessary for institutional confinement to be effective, co-operation of leprosy sufferers with British medical treatments was essential for British medicine to have any genuine claim to superiority over indigenous.

British claims to medical superiority extended beyond the clinic and were played out in the emergence of a scientific medical culture in nineteenth-century Britain and India and in the professionalization of British medicine. Leprosy research in colonial India contributed to the development of a scientific medicine in the colonies which was partly independent of Britain. Medicine at the 'periphery' was not entirely derivative from medicine at the 'centre', as has been suggested by MacLeod,[10] but facilitated its development. British and European research into leprosy in India contributed not only to the transformation of medical understanding in India but also to the professionalization of medicine in Britain.

Within India, from the 1870s leprosy and the control of its research and investigation became contentious ground in the struggle for power between the Indian Medical Service and its rival, the sanitary department. Not just professional power, but confinement itself was a negotiated condition. There was little agreement between levels of government, medical and legal authorities and Indian public opinion as to who among the leprosy sufferers was to be confined and the terms of confinement.

Further, while Foucault gives no place to resistance as a 'constitutive element in the history of power'.[11] in south India, resistance to confinement expressed both by British and Indian authorities, and the leprosy sufferers themselves, was essential to the formation and exercise of British authority. Far from being a unanimous articulation of power, confinement of leprosy sufferers was only ever partially endorsed by British and Indian authorities and was always dependent upon the co-operation of the leprosy sufferer.

The history of leprosy in India is part of the wider history of the poor. The vagrant Indian poor suffering from leprosy were the most visible to the British. The poor were the recipients of British institutional care, the test cases for new leprosy treatments and the targets of confinement. In Hindu culture it was the poor leprosy sufferer who was most affected by social and religious ostracism. The history of leprosy in nineteenth-century south India provides a lens through which the articulation of colonial power by medical and legal authority can be observed in all its complexity, ambiguity and limitation. Further, it gives some voice to those who have suffered leprosy in the past and an insight into the present experience of leprosy. It recognizes the suffering of disease as a significant part of human history.

1

Indian and British Concepts of Leprosy and of the Leprosy Sufferer

During the nineteenth century, British understanding of leprosy and of the Indian leprosy sufferer underwent considerable change. Humoral understandings of leprosy increasingly gave way to a more modern scientific concept of the disease and there were considerable changes in the British perception of the leprosy sufferer. In the early nineteenth-century, a leprosy sufferer was typically understood by the British in India to be Indian or Eurasian, male, poor and vagrant. What precisely constituted the disease of leprosy was still not clear. By the close of the century, the British had a coherent understanding of leprosy as a disease, and had come to recognize both the presence of leprosy in the Indian middle classes and the susceptibility of Europeans to the illness. Meanwhile, Indian understanding of leprosy and of the position of the leprosy sufferer had changed little. While the *dharmaśāstra*, the science of law, sacred duty and righteousness, placed restrictions on the leprosy sufferer's right to inherit and even prescribed outcasting in some circumstances, in practice most Indian leprosy sufferers continued in their family and working life, little hindered by the Hindu proscriptions.

Nineteenth-century indigenous and British understandings of leprosy

In the early nineteenth century, indigenous medical traditions accommodated a variety of medical understandings of leprosy. Siddha, the Tamil indigenous medical system, and Ayurveda, the Sanskrit system, recognized 18 types of leprosy, comprising seven forms of 'great leprosy', *mahākuṣṭha* (Skt), and 11 of 'slight leprosy', *kuṣṭha* (Skt), also broadly described as 'skin disease'. Psoriasis, leucoderma, syphilis and morphoea were categorized as milder forms of leprosy. In European medicine there was considerable change in the classification of leprosy during the nineteenth-century. In the sixteenth-century, the identification of syphilis as a condition separate from leprosy had caused a dramatic decline in the number of people identified as leprous. However, despite recognition that conditions described as leprosy on the basis of similar cutaneous appearance were not all the same disease, there had been little clarification of leprosy's distinguishing characteristics.

Throughout the eighteenth and early nineteenth centuries, leprosy continued to be a common term in European medicine for any disease in which the skin had a repulsive appearance, affected the whole body surface or was particularly resistant to treatment. However, as early as 1811, the distinctive characteristics of leprosy were already being identified in Madras. In 1811, James Dalton, surgeon at the Native Asylum, described leprosy in terms which were narrower than indigenous definitions of the disease and consistent with British understanding of the distinctive characteristics of leprosy developing by the mid-nineteenth century:

> The criterion by which leprosy indicates itself is by the skin becoming rough and scaly, a numbness in the hands and feet, the Voice is frequently hoarse, the breath very offensive, the lobes of the Ears are thickened and seem knotty and the Cheeks and the whole face are beset with large lumps of livid colour and it generally terminates in an Evil in which the fingers and toes gradually lose all sensation and become ulcerated and corroded, and at length the unfortunate sufferer finds himself daily falling to pieces.[1]

Although the bacterial cause of leprosy was not discovered until 1874, from 1842 publication of pioneering research by the Norwegian doctors C.W. Boeck and Daniel Cornelius Danielssen, a physician at St Jorgens Leprosy Hospital in Bergen, gave to European medicine a precise description of the disease. By studying the characteristics of *spedalskhed*, leprosy, in Bergen, Norway, where it was still endemic, and travelling extensively to compare *spedalskhed* with the same forms of disease in countries of the Mediterranean, Asia Minor and Egypt, Danielssen and Boeck were able to identify the characteristics of leprosy which distinguished it from psoriasis, syphilis and the other diseases often regarded as forms of leprosy. In 1847, they published *Om Spedalskhed* and *Atlas Colorié de Spedalskhed*, a two-volume work including post-mortem findings and clinical studies which became the Norwegian government's official report on leprosy for 1847. The Norwegians' work was best known in south India through the prize-winning French translation of this text published in 1848 as *Traité de la Spédalskhed ou Eléphantiasis des Grecs*.[2]

Ever since Danielssen and Boeck described leprosy as being of two distinct types, Nodular and Anaesthetic, the basic conceptualization of leprosy as existing in two distinct forms has remained central to descriptions of the disease. Current terms for the extreme forms of leprosy are lepromatous and tuberculoid. Between these extremes a spectrum of leprous forms, known as borderline leprosy, corresponds with the nineteenth century 'mixed' forms. The renowned Prussian pathologist, Professor Rudolf Virchow, described Danielssen and Boeck's work as 'the beginning of the biologic knowledge of leprosy'.[3] They were the first to describe leprosy as a distinct pathological condition, bringing the understanding of leprosy out of the humoral conceptualization of medicine which had persisted in Europe until the mid-nineteenth century, and into modern scientific conceptualization.

Although the correct nomenclature of leprosy continued to be debated in India throughout the century, by the mid-nineteenth century, Madras medical officers had developed a precise understanding of the disease, based on observation of leprosy sufferers under their care, which was consistent with the findings of Danielssen and Boeck. By 1860, leprosy was understood to be 'a constitutional malady, giving rise to certain morbid alterations in the sensation, colour, or thickness of the skin, and subsequently occasioning further lesions in the mucous membrane of the air passages, and even the deeper structures of the body'.[4] The British named the two major forms of leprosy 'anaesthetic' and 'tubercular', with the frequent co-existence of both forms in the same individual described as a 'mixed variety'.

An account of British understanding of leprosy in the Madras presidency by the 1860s is provided by the civil surgeon, Dr Francis Day, in his observations at the Palliport Lazaretto, Cochin, and the report on leprosy prepared in 1864 for the Leprosy Committee of the Royal College of Physicians by J. Shaw, Principal Inspector-General in the Madras Medical Department.[5] Shaw's report was based on information provided by medical officers directly involved in the care of leprosy sufferers. These British descriptions of leprosy not only indicate the extent of British medical understanding of the disease at this time, but also provide a grim picture of the physical devastation and suffering caused when leprosy was left untreated.

Lepra Anaesthetica

Day and Shaw noted that of the two varieties of leprosy, the anaesthetic form, *Lepra Anaesthetica, Poonah Kooshta Themir Coostarogum* [sic] (Tam), was the most common in south India but, in the early stages, often not recognized as leprosy. As described by nineteenth-century medical writers, the first indication of anaesthetic leprosy was a loss of sensation in hands or feet, or the development of localized patches of anaesthesia on the torso. After a few months, the anaesthesia extended from the extremities to the knees and elbows, and often further. Day reported a case in Cochin in which only the tongue remained sensitive to touch. The patches of anaesthesia tended to be small, irregular, lighter coloured than the surrounding skin, often glistening in appearance, and dry, because of the cessation of perspiration. Wrinkling or blistering of the skin over the extremities followed anaesthesia and the phalangeal joints became swollen and lost their flexibility. The blisters ruptured, leaving intractable ulcers which either healed slowly to be followed by other similar eruptions, or persisted, destroying the soft tissue of the hand to expose carious and dead bone.

While this ulceration was in progress, destructive changes occurred within the bones and joints. The joints became stiff and distorted, with interstitial absorption active in the phalanges. In the hand, the last, and often even the second, bone of the finger was absorbed, leaving nail and flesh at the end of the remaining phalanx. The same process of ulceration, absorption and foreshortening occurred in the feet, where ulcers often formed in the soles, corroding towards the metacarpal or carpal bones. As in the hand, these ulcers, which often appeared to have been cut out with a punch, eventually met dying or dead bone.

The British recognized that the destructive process of ulceration was often initiated by physical damage through burns or abrasions caused unwittingly because the hands and feet had become anaesthetized. Actions such as lifting a hot pot from the fire could cause severe burns in someone whose hands were insensible to pain. In addition, it was noted that vesicles which led to ulceration often formed on anaesthetized surfaces which had otherwise not been damaged. The British also observed the claw hand, characteristic of anaesthetic leprosy. In addition to ulceration and destruction of the bone in the hand, the fingers typically distorted and stiffened to become like bird talons, the first phalanges of the fingers bent backwards, while the second and last, curled and hardened into the shape of a claw. The hand, locked into this position, became wasted, withered and insensible.

Despite the severity of mutilation and the extent of anaesthesia which was possible, obvious deformity in the anaesthetic form of leprosy was usually confined to the fingers and toes, although damage to the nose, palate and eyes also occurred. Bleeding or, more frequently, a foetid discharge from the nostrils, signalled destruction of the bones of the nose and palate which made the voice hoarse and caused hearing loss. Ulceration of the mucuous membrane also occasionally occurred. Ulceration of the cornea and thickening of the conjunctiva were principal causes of blindness in leprosy sufferers.[6] The progress of blindness in untreated leprosy was acutely observed by Day in a patient who refused treatment at the Palliport Lazaretto:

> He was admitted into the Lazaretto in 1847; in 1859 he had lost the left eye, there was ozaena; and in January, 1860, a dark, purplish, painful swelling existed at the inner canthus of the right eye, just above the lacrymal duct, preventing the free passage of tears, which consequently ran over the lower lid, down the outer side of the nose. The lower lid was everted, the mucous membrane purplish and congested, the cornea hazy, in the centre of which, by degrees, a small opaque ulcer showed itself, along with passive congestion of the vessels of the conjunctiva covering the globe; this ulcer gradually enlarged, extending itself deeper, the anterior chamber was opened, the humours escaped, and sight was totally lost.[7]

Lepra Tuberculata

The second form of leprosy minutely described by Shaw and Day was *Lepra Tuberculata, Koostum Coostarogum* [sic] (Tam). As with the anaesthetic form, tubercular leprosy had no clear constitutional symptoms. The earliest signs of tubercular leprosy were a burning and itching sensation, sometimes felt in the face and extremities, with the skin becoming dry, as in anaesthetic leprosy, and bronze or fawn in colour. Raised patches appeared on the face and extremities and, in rare cases, on the torso. After an initial period of hypersensitivity, these patches gradually lost their sensitivity to touch. In the next stage of tubercular leprosy, nodules appeared on the nose, ears, brow and chin. In severe cases, the face became one mass of nodules and in some instances, the whole torso and extremities were similarly covered. The hair thinned and facial hair, including eyebrows and eyelashes, was lost. The nose became flattened, its aloe enlarged by tubercules, and the ear lobes became pendulous. This phase of the disease was often termed 'leonine' because the congestion of tubercules on the face, staring appearance of the eyes, loss of lashes and sallow complexion, gave the face a lion-like appearance. Tubercules also appeared on the tongue and mucuous membranes covering the hard palate, and discharges of blood and pus from the nose frequently occurred. As in anaesthetic leprosy, because of interference to the hard palate and nasal passages caused by the tubercules, the voice was altered or even entirely lost. The vomer, nasal and even, occasionally, the palatal bones, could become carious, completely destroying the nose. When tubercules invaded the upper end of the tear duct the same process of damage to the eye occurred as in the anaesthetic form of leprosy, resulting in complete blindness.

The British recognized that there was not the same degree of distortion in the hands and feet in tubercular leprosy as in the anaesthetic form. Although there was stiffening of the joints in tubercular leprosy, there was not the same tendency for the claw hand to develop and neither did interstitial absorption of the phalanges occur. It was possible for a leprosy sufferer in the most advanced stage of tubercular leprosy to have perfect hands. The damage to fingers and toes was caused, rather, by the third stage of tubercular leprosy, which was not always reached. In the third stage, the tubercules became irritable, frequently making the patient feverish, gradually breaking down to form foul and discharging circular ulcers, which 'may eat through a small joint', resulting in its necrosis and destruction. Ulcers did not only occur in areas of skin already thickened or affected by tubercules. Rather, as Day observed:

> perfectly circular [ulcers] often break out on the soles of the feet or palms of the hands, precluding walking and causing exquisite pain. Their appearance being as if a hole had been bored in an onion, showing its various layers throughout.

Although it was rare for joints to be entirely lost by those suffering tubercular leprosy, Day observed that, compared with anaesthetic leprosy, the final stages of tubercular leprosy are 'perhaps, still more loathsome to strangers and more disgusting to the poor patient whose body is often covered with discharging sores'.[8] Tubercular leprosy ran a swifter course and was more rapidly debilitating in its effect than the anaesthetic form of the disease which hardly seemed to shorten the leprosy sufferer's life. The distinct types of leprosy and their specific symptoms described by the British in the nineteenth century are all evident in untreated leprosy in India today.

As in indigenous categories of mild forms of leprosy, white leprosy or leucoderma, *veḷḷai kuṣṭam* (Tam), was regarded by some British medical officers as a third form of leprosy. However, Shaw argued that it was not leprosy but a form of albinoism, a type of alteration in skin pigmentation which, unlike leprosy, was 'without structural change, and without much functional derangement'.[9] This view prevailed and in the Government of India's later definition of leprosy for the purpose of census collection, leucoderma was specifically excluded. Developments in European classification of leprosy, together with direct observation, thus had a significant influence on nineteenth-century British medical understanding of the disease in south India. The precise description of the two principal forms of leprosy confirmed the growing recognition that leprosy was a disease distinct from syphilis and leucoderma.

It was evident to the Madras medical profession that leprosy sufferers did not die of leprosy but, rather, of other conditions, usually diarrhoea, dysentery, dropsy or bronchitis, to which they succumbed in their weakened physical condition. Shaw endorsed the opinion of Dr Rean, assistant surgeon at Chicacole, who reported that death among leprosy sufferers 'seems to be generally caused by want and the unfavourable influences to which they are exposed' but 'when well taken care of, they are capable of living to a considerable age'. Further, Shaw observed that those with leprosy were often also afflicted with other offensive skin diseases, including especially scabies, psoriasis, chronic eczema and venereal diseases.[10] Elephantiasis was also common among leprosy sufferers, especially those of the Malabar Coast. Partly because of the conventional use of the same generic term, 'elephantiasis', to describe both leprosy and elephantiasis, it was not until the early 1870s that it became clear that there was no physical connection between the two conditions despite their frequent co-existence.[11] Even with the medical treatment available today, leprosy sufferers are likely to die from infections and renal failure rather than leprosy itself, and are not spared skin cancers and other such conditions.

In the 1860s, before the understanding of germ theory and the action of bacteria, it was not understood that ulceration, principally caused by physical damage, initiated much of the bone deterioration in leprosy by communicating infection. By the early 1870s the presence in leprosy of granulation tissue 'deposited in the integuments, and in the fibrous textures, and about the nerves, and at times the nervous centres' was recognized. Although the bacterial cause of the granulations was not yet understood, the different distribution of the granulation, 'more plentifully, and particularly in the nerve structures of the body in the anaesthetic, and in the skin and mucuous surfaces in the tubercular form' was correctly recognized as explaining 'the peculiarities of the two forms'.[12] The leonine face often apparent in cases of tubercular leprosy, for example, was caused by the concentration of leprous bacteria in the face.

By 1874, Claude Bernard's experiments on the animal nervous system had led Deputy Surgeon General W. Johnston to state in the same year, quite correctly, that morbid changes in the nerves occurred 'long before any of the characteristic indications of leprosy have manifested themselves'. Johnston made a causal link between nervous damage and the external symptoms of leprosy by drawing on evidence of nervous involvement, implied in the Royal College of Physicians' report, and in Danielssen and Boeck's work. From their references to tingling and burning sensation along nerves in a limb, Johnston moved to demonstrate that:

> the local changes in the disease are manifestly results of alterations of nervous functions followed by organic lesions of the nerves supplying the external parts effected [sic] – the surfaces, the subjacent tissues and the bones becoming successively diseased, their sensitivity impaired or lost, and their structures disorganized.[13]

The role of the cutaneous nerves in causing leprous anaesthesia was noted also by Drs Lewis and Cunningham in their report *Leprosy in India* prepared for the sanitary department and published in 1877. This was a significant shift in emphasis from the view current in 1861 that leprosy was essentially a blood disease or dyscrasia, an understanding which had encouraged research into the agents involved in the formation of blood for evidence of leprosy causation.[14]

From the early nineteenth century, leprosy had been understood by the European scientific community to be transmitted principally by inheritance, a view held also by the Indian leprosy sufferers in Madras, though both medical officers and Indian leprosy sufferers also saw contagion as a possible means of transmission.[15] Danielssen and Boeck considered that leprosy was principally inherited, with about one-eighth of all cases attributable to 'spontaneous outbreaks'. Their scientific authority and the lack of any viable alternative, ensured that the hereditary theory of leprosy transmission remained the dominant medical view, even though the publication of *Om Spedalskhed* in 1847 caused scientific dissent. Some scientists advocated the miasmic theory of transmission and others suggested that leprosy was spread by contagion.[16]

The dissenters were not far wrong. By the mid-nineteenth century, the concept of bacterial disease causation was well advanced. As early as 1840, the fact that certain living organisms were the cause of contagious and infectious disease had been deduced by Henle, although it was not until the discoveries of Koch in the 1870s and 1880s that there was definite proof of the existence of bacteria. From the mid-nineteenth century, theories of disease transmission were increasingly confused and there was considerable tension between proponents of different scientific views. Despite the dominance of the hereditary theory of leprosy, as early as 1857 Norwegian research suggested that the disease was infectious, and Hansen was convinced that leprosy was contagious well before his discovery of the leprosy bacillus in 1873.

Hansen's discovery of the leprosy bacillus filtered into India quite quickly. Hansen sent scrapings from leprous nodules to H.V. Carter of the Bombay Medical Service, convinced that they contained the organism which caused leprosy. In 1883, Carter, who had become a principal authority on leprosy in India, was able to provide the Bombay government with his own illustrations of the bacilli discovered in a leprous nodule. Carter observed that, like syphilis and tuberculosis, the leprosy bacillus caused remarkably little local or general irritation in comparison with the violent symptomatic response to acute bacterial infection evident in cases of cholera. By 1883, European researchers had found the leprosy bacillus in the lymphatic system and blood, in diseased nerve trunks, the liver, spleen and testes: convincing evidence that leprosy was a systemic infection. The discovery of the leprosy bacillus did not, however, resolve the mystery of leprosy causation. Even though the British medical profession was confident that the pathology of leprosy was reasonably well understood, the means of propagating the disease was still unknown.[17]

The British anticipated that the origins of disease could be found in environmental factors, particularly climate, and diet. Some writers attributed the prevalence of leprosy on the Malabar Coast to humidity, and to consumption by the poorer classes of 'the flesh of enormous sharks and other coarse fish frequently in a state of putrescence'.[18] At the close of the century, the Leprosy Commission, implemented in 1890 to enquire into the state of leprosy in India, suggested, on the basis of census figures on leprosy distribution, that while there was insufficient evidence to endorse a theory of climatic causation, leprosy was least prevalent in dry areas.[19] The possibility of a link between climate and leprosy is still a matter of debate today.

Many writers did endorse the dietary theory, ascribing the origin of leprosy to the frequent consumption of stale or bad fish and of rancid oil and bad cereals. The theory that leprosy was caused by the consumption of fish had a long history and returned to favour in the late nineteenth century. Its most influential advocate was Jonathon Hutchinson who, at the Tenth International Medical Congress held in Berlin, argued strongly for a connection between the incidence of leprosy and the consumption of fish, a view confirmed in his 1906 publication *On Leprosy and Fish Eating*. T. Farquhar, a retired surgeon-major of the Bengal Medical Service, with Tilbury Fox, eminent dermatologist and physician in the Department of Skin Diseases at University College, London, played a substantial role in leprosy research. Farquhar discovered a link between leprosy and the consumption of inferior wheat grain in India, noting that leprosy was comparatively rare in districts where high-quality cultivation was a long-established tradition. Similarly, the consumption of various forms of dhal (lentil) was regarded as a possible cause of leprosy. Fox and Farquhar concluded, however, that leprosy may not be a consequence of the action of any particular aspect of diet but rather might result from the absence of particular factors in the diet, such as nitrogen and potash, and be exacerbated by poor housing conditions, bad hygiene, poor diet, fever and other such circumstances. They suggested, partly on the evidence of the Irish population which, though poor, was relatively free of leprosy, that the consumption of large quantities of potash, in the form of potatoes, could be a means of preventing the disease![20]

In linking leprosy with poverty, Fox and Farquhar endorsed the view held by the majority of medical officers working with leprosy in the Madras presidency. While medical officers did not specifically state that poverty was a cause of leprosy, it was identified as a condition in which leprosy was prevalent and which contributed to the full development of the disease. Madras medical authorities were unanimous in their belief that 'the dwellings of those afflicted with leprosy [were] generally extremely filthy and defective in all sanitary requirements'. Reflecting the link made in mid-nineteenth-century Britain between poverty and immorality, Shaw asserted that 'poverty, low living, hardship, filthy habits and debauchery aggravate and accelerate the disease' once it had become manifest.[21] Leaving aside any connection between poverty and immorality, leprosy is still recognized today as a disease exacerbated by poverty, poor hygiene and sanitation. The question of whether contagion or inheritance was the principal means of leprosy transmission and, more specifically, whether contact with a leprous ulcer or sexual intercourse with a leprosy sufferer were means of contagion,[22] was to remain unresolved until the close of the nineteenth century.

Despite much discussion and many advances in understanding the disease and its etiology, there had been little change in the essentially humoral British understanding of leprosy causation as late as the 1870s. As Lewis and Cunningham noted in their 1877 report for the sanitary department:

> with regard to our definite knowledge of its actual causation, it is to be feared that we have not, except phraseologically, advanced very much on the etiological views recorded by Atreya many centuries B.C., which were to the following effect: 'When the seven elements of the body become vitiated through the irritation of the wind, the bile and the phlegm, they affect the skin, the flesh, the spittle, and the other humours of the body. These seven are the causes respectively of the seven varieties of *Kushta*' leprosy.[23]

Although, in general, the humoral basis of British medical concepts was breaking down by the mid-nineteenth century, in the understanding of leprosy there was, as yet, nothing to replace it.

The understanding of leprosy among the Madras medical officers was entirely consistent with the broader colonial view. The other crown colonies agreed that leprosy was a constitutional malady which existed in two distinct forms, the anaesthetic and the tubercular, with both forms frequently occurring in the same individual. They experienced the same difficulties of correct diagnosis as the Madras medical authorities and believed in a close association between leprosy and poverty.[24]

British perceptions of the leprosy sufferer in nineteenth-century south India

It is very difficult to establish a true picture of the leprous population in nineteenth-century south India. Early-nineteenth-century British sources are almost exclusively concerned with the poor and vagrant leprosy sufferers who pressed themselves into the British line of sight, begging alms from British residents in the presidency towns or seeking medical care and shelter at the British-run Native Infirmary in Madras, a hospital for the Indian poor of every caste, completed in 1798. As a consequence, the British viewed leprosy as primarily a disease afflicting the poor. As A. Lorimer, secretary to the Madras Medical Board, commented in 1855:

> This disease is almost exclusively confined to the poorest classes of the community; the sufferings amongst them have ... been very great on account of the scarcity of food, and this may be looked upon as the principal cause of the increase of this disease, food not only in insufficient quantity but often of bad quality.

Later, in the Royal College of Physicians' report, Shaw recognized that leprosy 'undoubtedly attacks all races, European, East Indians, Musselmans, and Hindoos of all denominations, Brahmins, as well as Pariahs'. However, the notion of a link between leprosy and poverty, with the racial distribution of the disease roughly correlating with the relative economic status of the Indian, Eurasian and European population of Madras, remained. Shaw went on to say that the disease was 'rare among Europeans', with Eurasians more frequently attacked, though less so than the Indian population, 'especially the lower orders'.[25]

It was not until the first all-India census that the leprosy sufferer came to the notice of the British in the home and at work, as well as in the street and hospital or dispensary. Then it became evident that, while poverty was a factor in the incidence of leprosy, the apparently almost exclusive occurrence of the disease among poor Indians was in part a consequence of the tendency for those with leprosy among the higher socio-economic classes to conceal themselves. House-by-house enquiries made it clear to the British census officials that there was a large, concealed population with leprosy who were neither paupers nor vagrants. In addition to the process of census enumeration, the 1890 Leprosy Commission and discussions on the 1889 Lepers Bill and subsequent legislation, brought to both the Government of India and the local British officials first-hand knowledge of the demography and wide-ranging socio-economic status of the population with leprosy.

The tendency among both the British and Indians to conceal leprosy had a profound effect on the formation of a population profile of leprosy sufferers in the census. As H.H. Risley, census commissioner for India in 1901, explained:

[Leprosy] is regarded with horror and disgust and those who suffer from it conceal the fact whenever they are able to do so. It is commonly supposed that Europeans in India are practically exempt, but . . . this is by no means the case. The fact is generally not known because of the reticence of persons who suffer from it. In the case of natives of the country the same dislike to publicity is felt, but it is overcome amongst the very poor, who cannot otherwise support themselves, by the necessity of begging and of using their unfortunate plight as a means of exciting the compassion of the charitably disposed. In the case of those who do not beg, concealment is less easy amongst the poor than amongst the rich and amongst males, who mix freely with their fellow villagers and wear but scanty clothing, than amongst women, whose dress more completely covers them and who live a more secluded life; the prejudice against publicity is also much stronger in the case of women.[26]

Despite the exclusive focus on the leprosy sufferers of the vagrant and pauper class in the few early British records available, it is very likely that the population of the early nineteenth century was not much different from that of the later period. The person denoted 'leper' in early-nineteenth-century British government records belonged to the most visible group of leprosy sufferers identified by Risley. Usually male, he lived by begging on the streets of Madras and had signs of ulceration presumed to be leprous. He was part of the wider grouping described by the British as 'pauper infirm' and only survived by being visible on the streets. Poor leprosy sufferers, as the most visible and most in need of government assistance, remained the focus of government attention until the issue of legislation for confinement was raised in the late 1880s, provoking a more searching enquiry into the scope of leprosy in south India.

In Madras, quinquennial returns giving general population numbers by district had been prepared since 1851–52 by the Revenue Department. In 1871–72, the first attempt was made to take an all-India general census on a given date. The census included a category denoted 'infirmities' and a sub-section, 'leper'. The term 'infirm' indicated 'persons labouring under mental or bodily infirmity disqualifying them from earning a livelihood'. Surgeon-Major W.R. Cornish, sanitary commissioner for Madras, warned in his report on the 1871 Madras census that infirmity statistics were 'probably not very accurate'. He recognized that 'there is a reluctance with most people to admit the existence of physical defects and the figures therefore must be accepted with a liberal margin'.[27]

Throughout India, the actual identification of those who suffered from leprosy was highly problematic. In addition to the difficulties of dealing with different ways of identifying leprosy in the different languages and dialects, the terms used to describe some of the infirmities had different meanings in different areas of the same district. The census enumerators who gathered the raw census data were not specifically trained in physical examination but were unpaid Indian agents, selected by the government from the ranks of minor officials, community and religious leaders, local schoolteachers and students, for the sole purpose of gathering information and filling in the census forms on the appointed day.[28] Even in the late nineteenth century when leprosy had been well described in numerous scientific and medical journals and tracts in India and abroad, and there was fairly uniform British medical opinion on its symptoms, diagnosis of early leprosy was very difficult for trained medical practitioners and virtually impossible for untrained census enumerators. In its early stages, leprosy bore a striking resemblance to leucoderma, and confusion of leprosy with syphilis and various other skin diseases was frequent. Ulceration caused by other diseases was often mistakenly identified as leprous. The accurate diagnosis of leprosy was so difficult that the 1890–91 Indian Leprosy Commission found that even among those leprosy sufferers especially selected for its examination at various centres, as many as 9.5 per cent were suffering not from leprosy, but from diseases with similar external symptoms.[29]

The census enumerators could not even look to the leprosy sufferer for accurate information. In addition to those who deliberately hid the disease, there were those who were ignorant of their illness. H.A. Stuart, superintendent of census operations in Madras for 1891, noted in his report that a medical officer of the Madras General Hospital found it 'not uncommon to discover the presence of leprosy in a patient who had himself no idea of his unfortunate affliction'. Another reason for under-reporting was the difference between British and Indian perceptions of the physical condition of the leprosy sufferer. While the census classed 'Lepers' as one of the groups of infirm, the others being 'Insanes', 'Idiots', the 'Deaf and Dumb' and the 'Blind', some may have suffered from leprosy for years without being considered to be leprosy sufferers by the Indian community, much less considered to be 'infirm' according to the definition of the census.[30]

Instructions given to the census enumerators in the Madras presidency did little to alleviate the difficulties of correct diagnosis. In the 1871 Madras census, enumerators were simply requested to: 'state how many (if any) of the males and of the females are blind, deaf, dumb, idiots, insane, or lepers'. In the 1881 census, the enumerators were required to: 'Enter as "lepers" only those who are afflicted with the true or eating leprosy'. No description of the relevant identifying symptoms to fit this definition was offered. 'Eating leprosy' presumably referred to leprosy as manifest in signs of physical deformity, and could only describe the more advanced and often burnt-out, non-infectious cases of disease. Early signs of the disease, such as light-coloured patches of skin, could easily be missed, and other forms of ulceration which appeared to 'eat' the body were often mistakenly identified as leprous. Risley considered that the 1891 instructions were an attempt to reduce the misdiagnosis of early leprosy. He noted in his report on the 1901 India census that: 'it is probable that the efforts made in 1891 and at the present census to exclude cases of mere skin discolouration from the return have resulted in the omission of a certain amount of the true corrosive disease'. As a consequence, he expected that: 'The enumerators who had been specially warned against the entry of white leprosy, would thus be very likely to omit persons suffering from the initial stage of the disease.'[31]

The Madras census commissioners repeatedly commented on the inaccuracy of returns for leprosy in the presidency. Cornish observed in his 1871 Madras census report that the year's figures for all infirmities 'cannot be accepted as of much value' and pronounced the 1871 return of leprosy sufferers at nearly 96 000, or 1 in every 1875 of the population, to be unrealistically high. The 1891 returns suggested a radical reduction in cases of leprosy, only 12 617, or 1 in every 2824 of the population, being returned as 'lepers'. However, for the reasons stated above, Stuart observed in his report on the 1891 Madras census that this was a considerable underestimation of the true situation.[32]

Although the number of people suffering from leprosy in the Madras presidency cannot be accurately ascertained from the census, more can be said about the distribution of leprosy in the presidency. The 1891 census was the first to present a fairly full profile of the population of leprosy sufferers and to pay careful attention to details of distribution. The 1891 census endorsed the 1871 assertion that leprosy existed in every taluk of each district. Further, the 1871 finding of a higher incidence of leprosy in coastal districts than inland, was examined more closely in 1891. Specific coastal taluks of Ganjam, Vizagapatam, Nellore, Chingleput, South Arcot, Madura and South Canara showed a higher proportion of leprosy sufferers than the inland districts. In Godavari, Kistna, Tanjore, Tinnevelly and Malabar, however, the incidence was lower than in the inland areas. Further, the 1891 census found leprosy to be generally more common in dry districts with high seasonal rainfall.

In addition to a fuller examination of the 1871 distribution figures, the 1881 census presented a fairly full social and economic profile of the population of leprosy sufferers. The 1881 census built on the simple provision of age data in the 1871 census to suggest that the leprosy sufferer in late-nineteenth-century south India was usually between the ages of 20 and 59, male, of low caste and low socio-economic status and Christian. The 1891 age-statistics suggest that although leprosy afflicted all age groups, the incidence of the disease appeared to increase with age, more specifically, from 20 years up to the 55–59 age bracket.[33] The impression thus created, that leprosy tended to bypass the young in preference for those over 20 years old, was recognized in the 1901 Madras census as not being entirely correct. W. Francis, superintendent of the 1901 census for Madras and the Coorg, suggested in his report that the difficulty of early diagnosis was partly responsible for the paucity of cases of leprosy being reported among children. Risley noted in the same year that the tendency for leprosy to be latent in the body long before any physical symptoms appeared further contributed to the paucity of young people reported with the disease. The young were just as susceptible to leprosy as the elderly, but with late-nineteenth-century diagnostic methods the presence of the disease could not be identified. Contemporary surveys of leprosy in endemic areas of Tamil Nadu suggest that the incidence of leprosy peaks first among those aged 10–20 years and again in the older age group identified by the nineteenth-century commentators.[34]

Concealment of children with leprosy also resulted in under-reporting. Francis noted, in the 1901 Madras census, that even where the disease had become physically apparent, parents were reluctant to enter their children as afflicted with any infirmity 'as long as there remains any [sic] the remotest possibility of persuading themselves and others that the existence of the infirmity is still open to doubt'.[35] The British category of infirmity significantly overlapped with the categories of physical and mental disability which excluded a person from inheritance according to the traditional Hindu law as embodied in the *dharmaśāstra*, discussed below. Indian reluctance to admit that a child suffered an infirmity was not in deference to British cultural attitudes; rather, it reflected the concern within Indian society for children to be fit for full participation in religious, social and economic life.

The life expectancy of those with leprosy was also an area of contention. The sudden reduction in the number of leprosy sufferers in the 60-years-and-over age group recorded in the 1891 census, was repeated in the 1901 results for India. Commenting on the findings, Risley cited Danielssen and Boeck's estimates of the average duration of life from the onset of the disease, which were $9\frac{1}{2}$ years for tuberculated and $18\frac{1}{2}$ years for anaesthetic leprosy, and proposed that leprosy sufferers had only a comparatively short life. This suggestion was disputed, however, by Francis, reporting on the 1901 Madras census, who observed that leprosy 'does not seem to shorten life to any great extent'. Stuart, commenting on the Madras census for 1891, asserted that the higher incidence of leprosy reported among males than females could only partially be explained by concealment and indicated beyond doubt 'that males are far more liable to the disease than the other sex'. However, the question as to whether or not males are physiologically more vulnerable to leprosy than females is even today a matter of debate. Risley suggested in 1901 that the more public and social life enjoyed by men was a factor in their higher representation in the leprosy figures, and that cultural proscriptions against allowing women and girls a public life and, more specifically, contact with unknown men, tended to conceal from the census enumerators those who had contracted the disease. Women's fuller covering of clothing and restrictions on contact with the community are recognized today as factors reducing the risk of exposure to leprosy and accounting, in part, for the lower incidence of leprosy among women in India.[36]

The specific instructions given to census enumerators for the Madras 1871 and 1881 census illustrate well the difference in attitude within south Indian families towards giving information about females and males in the house. In 1871, the census enumerators were instructed that:

> the names of females are to be entered unless the head of the family objects. If he objects to providing the names, they must be designated by consecutive numbers. Thus if there are five females in the house, they will be No.1, No.2, No.3, No.4, No.5.[37]

In 1881, the situation was essentially the same. The enumerators were required to enter in full the names of all males and if given, the names of all females, 'but if there is any objection to telling the names of females, enter the word *"female"* in Column 2, and fill in all other particulars'.[38]

Francis made clear, in his report on the 1901 Madras census, the connection between female gender and concealment of leprosy in the census results:

> The enumerators had to enter the descriptions of the women which their male relatives gave, without them seeing them, and their relatives were not likely to readily admit that any of their sisters or daughters were afflicted with such an unpleasant disease as leprosy, though they might not mind stating that they were blind.

Further, he added, 'there ... appears to have been no more concealment of the existence of the disease among girls than among adult women'.[39] J. Chartres Molony, superintendent of the 1911 Madras census, saw family shame at the presence of an unmarried woman in the home, especially where leprosy had prevented marriage, as contributing to low reporting. He noted that: 'In India is found an especial reason for denial in the stigma which attaches to an unmarried woman and to her parents – a fact which may partially explain a marked sex disproportion in the leprosy returns of the last three enumerations'.[40]

Even by 1931, the census enumerators had not been able to resolve the problem of sex imbalance in the statistics. However, Hutton, reporting on the 1931 census of India, while noting the importance of the *purdah* system in concealing women and girls with leprosy, argued that although males were most likely to be exposed to the disease outside the home, they would also be a likely source of contagion for the females inside the home. The finding in the 1931 India census, that in the Hyderabad state there were almost as many women as men with leprosy among the small population of Christians, who did not practise *purdah*, confirmed that concealment was a major factor in the gender discrepancy in leprosy returns.[41] Such discussions of gender distribution of leprosy also raised the important issue of the social status of leprosy sufferers in Indian society. The presence of leprosy in a household could prevent a marriage, but, as will be discussed further below, in other circumstances leprosy sufferers themselves married freely.

Cohn has argued convincingly that the British believed the gathering of information on Indian caste and race through the census was the key to truly 'knowing' the Indian population, an activity considered necessary for the effective management of India. The gathering of information on the caste distinctions of leprosy sufferers implied that the British desired to know every aspect of the India they were struggling to hold, even if the information gathered was not entirely accurate. Risley reported in 1901 that caste analysis of infirmities had limited value. The all-India results suggested that lower castes suffered the most from leprosy. However, the commissioner warned that the practice of concealment was very likely to be a factor in the higher returns of leprosy among lower castes. He noted that although he had expected the practice of concealment to be greatest in the higher castes, 'the amount of concealment practised varies'.[42] However, such an expectation reflected more the British revulsion for leprosy and reticence about the presence of leprosy within their own ranks than Indian attitudes to the disease.

Again, contrary to British expectations, census reports indicated that the prevalence of leprosy in specific castes was related to the part of the country where the caste was based, not to any inherent caste characteristics or occupations. Further, within caste groups there was a wide divergence in the occurrence of leprosy. Endeavouring to find a reason for the caste distribution of leprosy, the 1901

Madras census report listed the various castes according to the material prosperity of the majority of their members. Confirming the finding of the Leprosy Commission, that the diffusion of leprosy was primarily dependent on the economic well-being of the people, the census superintendent found that leprosy did attack the poor and destitute more frequently than the rich and prosperous, 'while among the "rich" castes the number of lepers in every 100 000 persons of each sex is respectively 50 and 15, and among the "moderately wealthy" 51 and 15, among the "poor" castes it rises to 56 and 19.'[43] Even so, there were exceptions to this pattern. In a detailed investigation of leprosy in Chingleput District, surgeon Trimnell noted that leprosy existed among all the Indian races and castes, and appeared to be more prevalent among the rich than the poor, with the labouring classes more free from the disease than any other class.[44]

Further efforts to analyse the results were made in the 1901 census with tables added showing the average number afflicted with leprosy according to occupation and social-precedence groups. According to the tables, hill tribals and village menials were most likely to suffer from leprosy, with agricultural labourers the next most likely. Beggars and vagrants, despite being the most visible part of the population, were the least represented in the leprosy figures. This result was very probably because they were difficult to approach for census purposes.

By measuring the distribution of leprosy by social-precedence groups, the British probed further for a link between caste and the incidence of leprosy, reflecting the belief that caste was the key to Indian culture. The highest proportion of those afflicted by leprosy was to be found among the 'castes which eat beef but do not pollute except by touch', with the second highest proportion represented equally across three groups: the relatively ritually pure 'kshatriya and allied castes', *sudras*, who pollute by touch 'only to a slight degree', and the ritually unclean 'castes which eat beef and pollute without touching'.[45] Clearly there was no direct correlation between ritual impurity and vulnerability to leprosy. The incidence of leprosy among the more ritually impure was a consequence of low economic status rather than any custom or practice associated with low ritual status. Low-caste communities tended to be placed in a low socio-economic position which made them more vulnerable to leprosy.

Concealment of leprosy sufferers not only reduced the number of leprosy cases reported from higher castes, but influenced figures on the racial distribution of leprosy. Concealment of leprosy by European sufferers accounted largely for the belief that Europeans in India were 'practically exempt' from the disease. One option open to Europeans with leprosy, and not available to either Indians or Eurasians with the disease, was to return to England, a factor which probably also contributed to the very low incidence of leprosy reported among Europeans. In the years 1829–38 inclusive, only five civil Europeans, in all likelihood paupers and unable to return home, were admitted to the Presidency General Hospital suffering from leprosy. Even so, some British leprosy sufferers were not paupers and died in their own homes in south India.[46]

Eurasians, by contrast, figured highly in leprosy returns. The Friend in Need Society, established for the care of pauper Europeans and Eurasians in India, noted that among the 'classes of poor and distressed people' referred to them for relief, there were frequent cases of leprosy in those of 'Indo-British descent'. Some of those with leprosy seeking assistance from the Friend in Need Society had been dismissed from the Company's service and were driven by poverty to beg for their living. The president of the Friend in Need Society explained: 'as there is no provided asylum for them of any kind at Madras, and as no one will employ them as servants, they are necessarily thrown upon the Public, and become a burden and nuisance to the community at large'.[47] The fall of Eurasian leprosy sufferers from the security and status of Company employment attests to the capacity of leprosy not only to strike the poor but to relegate those of established socio-economic status to the ranks of the pauper class.

The signs of racial distribution of leprosy in the pauper class were consistent with the later findings of the Leprosy Commission developed on the basis of a fuller cross-class and cross-race analysis of the population with leprosy in the Madras presidency. The Leprosy Commission found that although 'there are no favoured races which are exempt from the disease', Indian peoples were the most prone to the disease and European peoples the least. Even so, the Commission emphasized that, as with caste, racial difference in the prevalence of leprosy 'may well depend on the inequality of social rather than racial conditions, the native ranking lowest in such a scale, while the Eurasian occupies an intermediate level'.[48] British recognition of the racial distribution of leprosy and admission that Europeans were as susceptible to the disease as Indians was remarkable in the climate of fear of European contagion of leprosy following the death of Fr Damien of Molokai in 1889. British attention to the racial distribution of leprosy was not only an aspect of the broader need to know the Indian population suggested by Cohn. The British had inherited the remnant of medieval European associations of leprosy with immorality, and possessed a greater revulsion for the disease than did the Indian population. Implicit in British attention to leprosy among Europeans was, thus, also a hope that the British in India could be distanced from leprosy.

In addition to the expectation that race was a factor in susceptibility to leprosy, the assumption that aspects of religious practice contributed to the disease was implicit in the census enquiries into the distribution of leprosy by religion. Though not explicitly stated, it is likely that the British expected the high incidence of leprosy among Indians to be connected to the Hindu religion which was increasingly represented by British Protestant missionaries as a religion which perpetrated poverty and 'bodily degradation'. However, the findings on caste in the census offered a corrective to such notions. Stuart, in the Madras census for 1891, stated that the highest incidence of leprosy occurred among Christians and the lowest among Muslims, as a consequence of the different distribution of the religions, rather than any aspect of religious practice which might predispose or protect from disease.[49]

Francis observed, in the 1901 Madras census, that the frequency of leprosy among Eurasians in comparison with members of any other caste or race in the Madras presidency accounted, in part, for the very high percentage of leprosy sufferers within Christianity. Further, in addition to emphasizing that there was little likelihood of a connection between Christianity and susceptibility to leprosy, L.S.S. O'Malley noted, in his report on the 1911 census of Bengal, Bihar, Orissa and Sikhim, that the 'fact that a number... [were] inmates of leper asylums, where they have been converted to Christianity' also contributed to the high representation of Christians among leprosy sufferers. In Madras, the conversion of leprosy sufferers was also occurring. In the Madras Leper Hospital for example, missionary attention from the mid-1870s had contributed to the numbers of leprosy sufferers embracing Christianity. By 1886, the Rev. G. Mackenzie Cobban of the Wesleyan Missionary Society had baptized 21 Hindus from the hospital and formed a small church there.[50]

The link between religion and leprosy implicit in the census report is to be expected. Although from the sixteenth century leprosy was no longer prevalent in Europe, the moral implications of the disease and the stigma associated with it persisted into the nine-teenth century. Thus, early-nineteenth-century British understandings of leprosy did carry the resonance of medieval European Christian notions of the disease. In 1811, Dalton's description of the decay and ulceration attendant on leprosy as an 'Evil' is a reflection of the tenacity of moral conceptualizations of disease causation in south India in the first part of the nineteenth century. There was a long European tradition of understanding leprosy as, at least in part, a moral condition, and as both a form of divine punishment and an opportunity for the exercise of Christian charity. By the 1860s, when leprosy was clearly identified by the British as a constitutional dis-ease, the medieval European belief that leprosy was a consequence of sin had lost ground.

The link between sexual immorality and leprosy; the belief that leprosy also caused a voracious and malicious sexual appetite, was, however, slower to decline, despite evidence to the contrary by those who observed the disease directly. In 1824 the *Lancet* offered a cor-rection, printing the opinion of an indigo planter from the district of Tirrhoot. The planter noted that a person 'attacked with the species of leprosy prevalent in India' has:

> no excessive propensity to venery after the disease appears, al-though they may have had it before. It is a common opinion, that people seized with this malady are of a warm and amorous temperament; but, when a person is seized with leprosy, the plea-sure derived from such indulgences, and the capacity for them, are in a great measure annulled.[51]

However, among some British medical officers, writing in the 1860s, *'libido inexplebilis'*, insatiable desire, was still believed to be charac-teristic of leprosy.[52] The tenacity of this notion may be explained by the anxiety of racial mixing which attended the spread of empire, combined with the British perception that leprosy was principally a disease of the vagrant Indian male who lived without restraint. To this was linked the lingering belief that leprosy was transmitted sexually in the same way as syphilis and, by implication, that a leprosy sufferer was likely to have had many sexual partners. The British tended to be at some pains to disassociate from such behaviour any of their own number who contracted leprosy. The Medical Board Certificate stating that John Reginald Hand, deputy magistrate and collector in the revenue department, was permanently incapacitated for service as a result of leprosy, added: 'His incapacity does not appear to us to have been caused by any irregular or intemperate habits'.[53]

The spiritual resonance of leprosy regained prominence in late-nineteenth-century British thinking when the medical treatment and care of leprosy sufferers became increasingly the province of Christian missionaries. Hansen's discovery of leprosy as a precise disease, in which the biblical experience of leprosy could be recognized, coincided with the evangelical revival in Britain and the civilizing mission of the expanding empire, providing rich soil for the growth of Wellesley Bailey's vision for missionary care of leprosy sufferers in India. Bailey's Mission to Lepers in India and the East was founded in 1874 and, from small beginnings at Ambala in the Punjab, by the mid-1880s was assisting Christian individuals and missions to provide for leprosy sufferers in south India.[54] Partly as a consequence of the Lepers Act, the Mission increasingly took on responsibility for the care of leprosy sufferers in India from the turn of the century.

Until the beginnings of substantial missionary involvement in leprosy care, the spiritual aspect of leprosy had remained at the periphery of the medical officers' work, being principally the concern of chaplains who attended the leprosy hospitals. Gussow argues convincingly that the perception of the physical corruption of leprosy as symbolizing the spiritual corruption of the soul, which characterized medieval European concepts of leprosy, was revived with the restoration of a 'parabolized moral status' to leprosy sufferers in the context of missionary medical care. As Bailley exclaimed when confronted with leprosy in India: 'What a picture of the poor sinner does the outcast leper present!'[55] Not that the evangelical missionaries neglected the leprosy sufferer's physical care in preference for the spiritual. Provision of the best possible facilities for physical care and medical treatment was always essential to the Mission's work.

From the missionary point of view, however, even when clinical medicine could still do little, at least the leprous soul, often damaged not only by sin but also by the suffering of social ostracism, could be cared for and perhaps eventually healed.[56] Although they were the most numerous and influential, evangelical missionaries were not the only ones who provided medical and spiritual care for leprosy sufferers. At the Roman Catholic Jesuit Leprosy Hospital at Mangalore, whatever the perception of the leprosy sufferer as a spiritual being, efforts to find a physical cure also remained a priority.[57]

The position of the leprosy sufferer in Hindu culture

Although missionaries were important in leprosy care towards the end of the nineteenth century, throughout the century, British medical officers viewed the disease as primarily a physical condition. In the Hindu tradition, however, which was closer to the culture of medieval Europe in which religion pervaded society and the practice of medicine and law, leprosy was not simply a physical sickness, but the manifestation of the sufferer's spiritual condition. The Hindu *śāstric* tradition identified leprosy as a condition entailing profound ritual pollution, a state which had considerable implications for the sufferer's relationships with family and community. *Śāstric* authorities allowed for the outcasting of leprosy sufferers displaying ulceration. Those who were outcaste lost their ritual and social identity and thereby their capacity to inherit from their family. Even so, *śāstric* tradition did not neglect the outcaste sufferer completely, but required that the family provide maintenance for their physical support. Those not exhibiting ulceration were not susceptible to outcasting and thus could retain their inheritance rights.[58]

By contrast, there was no concept of outcasting or impediment to inheritance imposed on leprosy sufferers under Moslem law. The Islamic law of the Hanafi School, founded in the eighth century, was introduced into India by the Mughals and, as with the *dharmaśāstra*, Moslem civil law continued to be applied to Moslems during British rule. Under Moslem law, leprosy sufferers were not excluded from inheritance nor were the blind, insane and mentally deranged, groups all excluded by *śāstric* law. Like Hindu law, however, Indian Islamic law prescribed maintenance for an adult infirm relative.[59]

Śāstric law, however, did not apply equally to all Hindus who suffered from leprosy. Though theoretically applicable to the 'twice born',[60] the *dharmaśāstra* was primarily relevant to the Brahmins and was virtually ignored by the non-Brahmin and tribal peoples. In south India, the Brahminic influence was generally far less than in the north, and the extent to which even the *Mitākṣarā* the principal text of the southern school of Hindu law, could be applied in the south was extremely limited. Traditionally, the textual *dharmaśāstra* would have been followed by non-Brahmins only where it correlated with custom and even among Brahmins it would often be superseded by custom. Most legal disputes among Hindus were settled according to customary methods which depended upon the caste, social status and location of the parties involved and often did not include consultation with *śāstric* pundits.[61] Where leprosy was a factor, as W.A. Willock, collector of Vizagapatam, explained in 1889: 'the enforcement of the disabilities imposed by Hindu law on the more aggravated form of the disease depends on the social position and wealth of the sufferer'.[62]

Outcasting was certainly not imposed on all leprosy sufferers with the ulcerous form of the disease. The 1911 census recorded more than 12 000 leprosy sufferers as married or widowed, 8500 of whom were married. Many of these marriages were no doubt contracted before the disease became obvious. However, in the age group 1–10, 12 boys and one girl were married as were 13 boys and 11 girls in the age group 10–15. As Willock noted, if the leprosy sufferer was wealthy enough, 'his affliction interferes but little with his family arrangements'.[63]

Aversion to the leprosy sufferer appeared to be greater among the British in India than among Indians, despite the *śāstric* injunctions against leprosy sufferers. As J. Thompson, the collector of Coimbatore, commented in 1889: the 'degree . . . of [leprosy's] repulsiveness to the healthy individual' was 'extreme, I think, only or mainly to the European'. Certainly Willock was far less accepting of marriage by leprosy sufferers than was Indian society, noting with regret that in his district there were 'miserable cases of girls being married to men known to have the disease in a well-marked form'.[64]

This freedom of interaction, so abhorrent to the British, was evidently of less concern to the Indian population. Sufficient people were prepared to do business with leprosy sufferers to enable them to earn their living publicly. Surgeon-Major A.J. Laing, sanitary commissioner for Madras, observed in 1889 that leprosy sufferers were, in practice, allowed to mingle 'unchecked . . . with the ordinary population of the country'. Those with leprosy often worked in markets and bazaars and went about freely in public. Further, a small proportion of leprosy sufferers were literate, a number probably commensurate with the rate of literacy in the Indian population as a whole, and earned their living using these skills. Of a total population of 12 674 males and 4184 females with leprosy, the 1891 census recorded 1700 men and 45 women as literate, probably in their own vernacular rather than in English.[65]

Both literate and illiterate leprosy sufferers were able to earn their living in contact with the public. Le Fanu, collector and district magistrate of North Arcot, recalled in 1889 that both the baker and chief butcher at his station were afflicted with leprosy. Further, he recalled the appearance of a *vakil* before him in court 'in an advanced stage of leprosy, with matter running from open sores'. Willock noted barbers, merchants, a sweet maker, a shepherd and a goldsmith all suffering from leprosy and all working in the Vizagapatam District. Further, a brahmin schoolmaster continued to teach despite evident signs of the disease. The death of his brother and sister from leprosy, and the infection of his wife and three children did not prevent him continuing to earn his livelihood. The chairman of the Municipal Council, Cochin, noted that in Calicut and many other places, leprous persons were visible in town 'carrying about, and exposing for sale in petty shops kept by them, articles of food and drink and other miscellaneous goods'.[66]

The 1911 census offered a detailed breakdown of the occupations followed by leprosy sufferers. Of the number counted, both male and female, totalling 16 858, nearly 6000 were recorded as engaged in cultivation, 155 raised farm stock and thus also supplied milk, ghee and cheese; 281 were connected with 'industries of dress and toilet;' 59 with 'food industries' and 422 with 'trade in foodstuffs'; 14 male leprosy sufferers were counted as restaurant or hotel keepers; 17 practised medicine; 33 were in domestic service; 38 were fishermen and probably sold their catches; 38 were school masters; 11 were policemen; 77 were engaged in religious duties and six were employed in 'public administration'. Of the total number of 16 858 there were 9707 unaccounted for, suggesting that the majority of leprosy sufferers were either financially supported in the home or lived by begging.

Though not mentioned in the census, there were also small numbers of leprosy sufferers admitted to hospital from both the Native and European regiments of the army.[67] British policy concerning the fate of recruits who contracted the disease is unclear. However, the admission of leprosy sufferers to hospital and the treatment in the 1870s of a private from the Native Infantry by fumigation with carbolic vapour,[68] suggests that, from the Native regiments at least, dismissal was not attendant on contracting the disease. For leprosy sufferers in the Civil Service the situation was different. British fear of leprosy was evident in the revised *Civil Medical Code* for Madras which declared that 'the mere existence' of leprosy was 'a sufficient disqualification for the public service'. It was not until the *Code*'s fourth edition, published in 1929, that this ruling was altered to provide the opportunity for a government servant with signs of leprosy to receive treatment. Under the new ruling, the government servant would be invalided from service 'only if after undergoing this treatment for the full period of leave to his credit he is still certified to be infected with the disease'.[69]

The change in stance was probably a consequence of advances in leprosy treatment, particularly the relative success of chaulmugra oil by the early twentieth century. Although in the early 1870s, gurjon oil and other treatments had found favour, British hopes for their success were relatively short lived. The confusion surrounding the contagiousness of leprosy was also possibly a factor in the dismissal of civil servants. However, that a member of a Native regiment could receive treatment for leprosy before the mid-1870s, whilst at the same time a civil servant with leprosy faced instant dismissal, suggests that medical considerations were not the only factors influencing the position of leprosy sufferers in the Indian Civil Service. British fear of close contact with leprosy in the work place, and reluctance to recognize the susceptibility to leprosy of Europeans, who in the 1870s were still considered to be largely exempt from the disease, were also likely to have been influential. Dismissal of members of the Civil Service with leprosy also suggests a British concern to preserve institutional authority by ensuring that the public face of British rule remained untarnished by offensive disease. Indian officials representing British authority were thus as susceptible as the British to such dismissal.

Dismissals did not necessarily run smoothly, however. In 1890, Subramanya Chettiar, a local bench magistrate, was requested to resign on the grounds that he was suffering from leprosy. Subramanya Chettiar, however, denied that he had the disease and refused to stand down. The Madras Judicial Department gave him the option of appearing before a medical board or having his powers as special magistrate withdrawn. Although the outcome of Subramanya Chettiar's case is unknown, mistakes in dismissals for leprosy did occur. In 1893, the first clerk of the sub-registrar's office, Cherpalcherri, Calicut, was declared to be a leprosy sufferer and 'permanently incapacitated for further service'. He appealed to the Madras Civil Medical Board which pronounced him free of leprosy in March 1894. The clerk was reinstated the following month and claimed full pay for the time that he was absent from employment. The Government of India granted his request as a special case.[70]

The Indian attitude to those leprosy sufferers who lived by begging was, however, similar to the British horror of the disease. As Willock observed: 'the unfortunate victims whom we see begging about the streets . . . are shunned and avoided by everyone'.[71] As Day recognized in 1860, economic status was critical to the leprosy sufferer's social position in Madras:

> The leper in this part of India, as long as he can work, or has money, is no outcast from the society of his fellow creatures, who live in the same house, partake of the same food, or even intermarry with him; but when the time arrives that he is unable to work, he becomes uncared for by all.[72]

In Madras at least, it was not customary to care for the poor or dying leprosy sufferer.

Vagrant leprosy sufferers were allowed access to drinking water in ponds and water reservoirs, and mixed in the large crowds at great festivals and other public occasions.[73] However, contact with the public was not as close as in the exchange of goods and services, and it was the vagrant and poor leprosy sufferers who most often needed and sought shelter and medical care in the institutions administered by the British.

While British medical understanding of leprosy became more scientific during the nineteenth century, leprosy continued to be susceptible to moral and spiritual associations, and to inspire revulsion and fear among the British. In Hindu culture, the spiritual significance of leprosy and its consequences for inheritance and social status were expressed to some extent through the *dharmaśāstra*. However, in practice it was socio-economic status which had the most profound effect on the leprosy sufferer's place in Indian society.

2
Patient or Prisoner?
Leprosy Sufferers in British Institutional Care

Since the publication of Foucault's *Madness and Civilization* (1961) and *Discipline and Punish* (1975), leprosy, poverty, criminality and insanity have tended to be understood less as discrete conditions and more as characteristics which unite those living 'in the margins of the community' and mark them out for exclusion.[1] In nineteenth-century colonial south India, the leprosy sufferer carried not only the ancient stigma of leprosy but also, to varying degrees, the nineteenth-century stigmas of vagrancy, poverty and criminality. Although leprosy affected all races and socio-economic groups, throughout the nineteenth century it was the poor and vagrant leprosy sufferers, the vast majority of whom were Indian and Eurasian, who were the focus of British attention and were subject to the imposition of British legal and medical authority. Even so, the leprosy sufferer was by no means a passive subject of British power. The leprosy sufferer's capacity to resist confinement and to contribute to the different levels of British medical, legal and government authority was far greater than that admitted by Foucault.

Until the Government of India proposed, in 1889, an act to confine leprous vagrants, much of the initiative for confinement in southern India was from local medical and charitable authorities, and involved disagreement and negotiation with both presidency governments and the Government of India. The relationship between poverty, criminality and leprosy was complex, and those in power had very different perceptions of the leprosy sufferer. The sufferer usually first experienced the links between poverty, leprosy and confinement developing in south India through police intervention and hospitalization. For many, their ambiguous status in the institutional context – the confusion as to whether they were more truly patients or prisoners – became clear only when they tried to leave the hospital.

Institutions

Institutional care for leprosy sufferers was slow to develop throughout the nineteenth century. The East India Company gave first priority to the health of the European soldiers and second to the health of the tiny European civilian community. In Madras, the East India Company established the first hospital for the care of Europeans in 1664 and opened a Naval Hospital in 1745. The first hospital for the care of the Indian sick was opened in 1799, more than fifty years later.[2]

The oldest institution in the Madras presidency specifically for the care of leprosy sufferers was the Palliport Lazaretto at Cochin. Said to have been established by the Portuguese in 1587, it was restored by the Dutch in 1728 and continued by the East India Company as a leprosy hospital in 1795. The Lazaretto was initially maintained with rent money from the river ferry which ran from Vypeen to Cochin. When this became inadequate the Madras government provided each leprosy sufferer with clothing and bedding and a daily allowance of one *ānna* ($\frac{1}{16}$ rupee), six *pie* ($\frac{1}{192}$ rupee).[3]

In Madras City, the first substantial shelter for leprosy sufferers was offered by the Monegar Choultry and Native Infirmary and, in 1816, the Madras Leper Hospital built under its patronage took on the role of leprosy care. The Madras Leper Hospital developed on an *ad hoc* basis as a result of crisis and pressure rather than any coherent government policy, and was administratively separated from the Monegar Choultry and Native Infirmary only in 1840 when it came under direct government patronage. Leprosy sufferers received care and shelter at the Monegar Choultry and Native Infirmary,[4] two separate charitable institutions established by the British in the late eighteenth century which were combined in 1809.[5] The choultry's object was principally to give relief and shelter to the poor, particularly 'the Lower Orders of the Natives' but also to enable the Madras government to clear the streets of vagrants by providing shelter for the destitute.[6]

In 1813 several factors led the Monegar Choultry's management committee to recommend that the Madras government build a separate hospital for leprosy sufferers. The increase in the number of leprosy sufferers seeking admission, largely in the hope of being cured by Surgeon Dalton's treatment, discussed below, was placing considerable strain on tight funds. Further, the poor who were housed with the leprosy sufferers and the servants working at the Monegar Choultry were 'continually stating their fears, of being infected themselves, and their children, with the distemper'. Repeating the appeal four months later, the management committee added that, despite all possible care being taken 'to preserve cleanliness', the choultry was 'becoming offensive, from the number of sick persons ... many of them in the worst possible state of disease'. Convinced that Dalton's cure was ineffective, the Madras government, however, requested that the management committee make clear to leprosy sufferers resident at the choultry that their hopes were unfounded and encouraged their return home. Short-term temporary accommodation was provided for those who wished to remain but any new admission of leprosy sufferers was prohibited.[7]

The Madras government's measures did relieve the pressure on accommodation at the Monegar Choultry and Native Infirmary but not the pressure on the consciences and better feelings of those working at the asylum who wanted to assist the leprosy sufferer. In August 1815, the committee sent an impassioned plea to the new governor, Hugh Elliot, requesting sanction for accommodation specifically for leprosy sufferers and allocation of funds, either from the charity or some other government-nominated source, to feed, clothe and give medical care to any leprosy sufferers who came. In place of the temporary building, the committee hoped to buy or rent a small piece of land in the vicinity of the choultry and to build a designated bungalow of permanent materials. The committee argued that the government's prohibition on admission of leprosy sufferers in 1813 had been devastating in its consequences. Those discharged from the choultry and infirmary had 'made their appearance in the public wards and streets in the town and vicinity of Madras to beg alms for their relief, being reduced to the same misery they were suffering under before their admittance into the Infirmary'. The Monegar Choultry presented the leprosy sufferers as victims of the government's actions, 'helpless objects' labouring under 'very great hardship', and appealed to Elliot to act as the 'humane and benevolent' source of their relief.[8]

The committee's persistence, and possibly the change in governor, yielded results. In 1816, the first permanent, government endorsed and financially supported shelter and hospital specifically for leprosy sufferers of any race was built at the Monegar Choultry and Native Infirmary. The shelter was exclusively for men with leprosy. Women and girls with the disease continued to be housed in the hospital with the other patients.[9] However, the great need for leprosy care in the Madras presidency and the problem of overcrowding was not fully resolved. As Surgeon James Lawder, in charge of the Native Infirmary, commented:

> Under any circumstances an overcrowded hospital is highly objectionable, and if such is the case with persons labouring under ordinary complaints, what must it be to persons affected with the loathesome disease of leprosy.[10]

Pressure on the Madras government to increase accommodation for leprosy sufferers came also from the Friend in Need Society, the oldest charitable institution in Madras, which was dedicated to the relief of poverty and suppression of begging among the Europeans and Eurasians of Madras city. The Friend in Need Society requested that the government make arrangements for Eurasian leprosy sufferers to be accommodated separately from the Indian patients, thus catering to their need for a more European diet and housing. A further expansion of the leprosy hospital at the Monegar Choultry and Native Infirmary was sanctioned and, in July 1840, financial responsibility for patients in the leprosy hospital was completely taken over by the Madras government.[11]

While the specific needs of Eurasian leprosy sufferers did provide an impetus to the government in considering an extension to the leprosy hospital, ultimately expansion was sanctioned on the grounds of a general need for more accommodation. On further investigation, the Madras government found that there was no evidence of an increase in Eurasian leprosy sufferers in Madras and that the numbers of Eurasians afflicted with leprosy in the presidency had never been substantial. Even so, the Madras government decided that in the new leprosy hospital, Eurasians would 'be treated in every respect as Europeans' and would be accommodated separately from the Indian patients.[12] Leprosy evidently did not dissolve distinctions of race either for the sufferer or for the British.

Although concepts of segregation and hygiene as measures to prevent the spread of leprosy were not fully developed until the late nineteenth century, the ideas were already apparent in the specifications for the new Madras leprosy hospital. Prior to the building of the new leprosy hospital, with its separate cook house, store room and rooms for a nurse and assistant apothecary, it had been impossible to keep hospital clothing and other items distinct from those of the other sections of the choultry. As Lawder noted, 'the constant intercourse that unavoidably exists between the Leper and other Patients, (chiefly by means of the servants)' could not be prevented. Hospital and infirmary clothing was washed in common by the same person, and food for the leprosy hospital was cooked in common with that for the infirmary patients and served out to all by the same servants. Lawder argued that in the new leprosy hospital it would be necessary to keep 'Leper Patients, their attendants and every thing connected with them, ... perfectly distinct, and unconnected with all classes of Patients' [sic].[13]

The effort made to cater specifically for the leprosy patient's needs and the idea of hygiene as an aspect of their care were evident in the detailed provision of furniture for the hospital's extensions. Lawder recommended the use of iron cots overlaid with closely fitting loose wooden boards to make them more comfortable since other types of cot became infested with insects and were impossible to keep clean. Attention to patient's needs and hygiene remained a feature of the hospital throughout the nineteenth century. The 1895 standing orders noted that on admission every European patient would be supplied with clean clothes and bedding, and each Indian patient would receive a clean quilt and body cloth. All bedding and clothing was to be changed once a week or more frequently if necessary.[14]

In terms of practical compassion the Madras government's provision for pauper leprosy sufferers was, by the mid-nineteenth century, at least on a par with, and in many instances superior to, that available in other crown colonies. Any sort of establishment for the leprosy sufferer's care was rare, and where hospitals and asylums did exist, there was great variation in conditions and arrangements for accommodation. In 1864, there were three Lazarettos in the Madras presidency, two government asylums, one each at Madras and Cochin, and a private asylum run by the Jesuits at Mangalore.[15] The Madras presidency was more advanced than Bombay or Calcutta in the quality of leprosy care available. By the 1850s in the Madras Leper Hospital:

The patients were classified according to their previous habits and position in life. Books were provided for the educated, and gardening and other light occupations conducive to health and happiness were encouraged.[16]

Leprosy sufferers also received out-patient and, in some areas, in-patient medical care in the civil hospitals and dispensaries of the Madras presidency. Hospitals for both European and Indian sick existed in the Madras presidency from before the nineteenth century and there were dispensaries in the city from as early as 1829. In addition, in 1842, the Madras government formally established the expansion of a civil hospital and dispensary system throughout the presidency, with the first institutions to be opened in Trichinopoly, Madura, Masulipatam, Nellore, Bellary and Cuddapah. Dispensaries were open to patients who had suffered an accident, required surgery, had persistent ulcers which resisted cure by Indian medicine, had a severe or unusual disease, or were sick or diseased paupers. Patients with ordinary diseases, such as fevers, readily treated by practitioners of Indian medicine, or with incurable or untreatable diseases, were not to be admitted. Patients could attend the dispensary as out-patients, and both medicine and medical advice were to be provided at specific hours.[17]

Despite the stipulation that those suffering an incurable disease were not to be admitted to civil hospitals or dispensaries, the policy of admitting the sick poor and those with persistent ulcers enabled leprosy sufferers to receive care as in-patients. The number of leprosy sufferers receiving care in civil hospitals and dispensaries was very small but significant. In 1878 for example, 167 Madras civil dispensaries treated a total of 2172 leprosy sufferers as out-patients and 850 as in-patients, 60 of whom died. Of these, eight were out-patients and 487 were in-patients at the Madras Leper Hospital, and 43 were in-patients at the Palliport Lazaretto. The total number of patients treated in these dispensaries during 1878 was 947 132. In some civil dispensaries specific provision was made for in-patient care of leprosy sufferers. Chingleput Dispensary, for example, opened in June 1850, and nine years later completed two new wards designated for leprosy sufferers. Returns for civil hospitals and dispensaries were far more accurate than census statistics with returns from larger institutions regarded by contemporaries as completely correct.[18]

In south India, a few leprosy sufferers also received care from *langerkhanas* and *chuttrams,* traditional indigenous charitable institutions, some of which received government funding in the nineteenth century. The *langerkhana* was a free kitchen and alms house, maintained by public or private funds, which provided food and alms to the poor. The *chuttram* was a south Indian house where pilgrims and travellers of the higher castes could be fed and sheltered for a few days. Some were exclusively for Brahmins, and others provided for all castes. The *langerkhanas* for which figures are available indicate that most gave assistance to only one or two leprosy sufferers a year. The Vellore *langerkhana* was notable for providing relief for seven, eight and nine leprosy sufferers in the years 1885, 1886 and 1887 respectively. *Chuttrams*, in Tanjore District at least, provided for leprosy sufferers and others requiring relief. In Tinnevelly District there were 12 *chuttrams* with local funding and 69 with private funding. No separate financial records were kept for leprosy sufferers, but it is evident that facilities were available for their admission separately from the other poor.

Some small local hospitals also made provision for leprosy sufferers. The Local Fund Hospital at Chittor in North Arcot District provided a detached building with four beds for in-patient care of leprosy sufferers. From 1895 to 1897 the hospital assisted 152 patients of whom 108 were out-patients. The Poor Houses at Salem and Vizagapatam, financed from municipal funds, provided shelter to a total of 63 leprosy sufferers from 1895–97. Cochin Poor House, a private institution, did not admit leprosy sufferers, presumably because of the proximity of the Palliport Lazaretto and the mission-run Calicut Leper Asylum, established in 1894. Certainly, there was a need for leprosy care in Cochin. The Palliport Lazaretto treated 161 patients from 1895 to 1897 and the Calicut Asylum treated 164.[19]

Affliction with leprosy did not preclude other afflictions managed with institutional care. Leprosy sufferers were also institutionalized in the 'Insane and Idiot' asylums of the Madras presidency. In 1845, of 179 cases of illness in the Idiot Asylum attached to the Monegar Choultry, there was one case of leprosy. In the Waltair Asylum for 'lunatics', measures were taken in 1891 to segregate two residents once they began to show symptoms of leprosy, and orders were obtained from the government for their removal to the Madras Leper Hospital.[20] The growing awareness of the contagiousness of leprosy from the mid-1870s contributed to the continued efforts to segregate leprosy sufferers from other patients in institutional care which had begun at the Madras Leper Hospital in 1840.

Leprosy sufferers were also to be found in the prisons of the Madras presidency. Until 1840, when provision was made for a hospital within the Justice's jail, with another to be constructed in the House of Correction, sick Indian prisoners were sent to the Monegar Choultry and Native Infirmary for Treatment.[21] From 1829 to 1838 there were nine hospital admissions and one death from leprosy in the Madras Black Town Jails. In 1889, a specific ward was established at the Madras Leper Hospital for prisoners with leprosy.[22]

Leprosy sufferers in institutional care

The vast majority of those leprosy sufferers housed in the Monegar Choultry and Native Infirmary and, from 1816, in the Leper Hospital established there, had not been transferred from other institutions. The basis for the predominant British image of the leprosy sufferer as poor, male and Indian, could be seen in the population breakdown of leprosy sufferers in the Native Infirmary between 1812 and 1813: The youngest was 12 years old and the oldest 80 years, with the majority aged between 25 and 50 years. There were only five females to 86 males.

The population of leprosy sufferers in the Native Infirmary also reflected the later census findings that leprosy afflicted all without regard for religion or caste. Among the patients were Moslems and Hindus and, while a substantial portion of the patients were pariahs, higher-caste nairs were also resident. Although leprosy sufferers were driven by poverty to seek care at poor houses and *langarkhanas*, not all those housed in the leprosy hospitals were vagrants, though many were engaged in poorly paid work. The occupations followed by those in the Madras and Cochin Leprosy Hospitals indicated that the majority were employed as labourers and cultivators, with cooks, fishermen and artisans substantially represented. Only slightly more than 9 per cent were recorded as unemployed or beggars.

The admission/discharge profile of the Madras Infirmary's population of leprosy sufferers suggests that although some sought long-term shelter at the infirmary, most tended to remain there for only three or four months until their condition improved, then returned to work and families. Of the 86 leprosy sufferers in the infirmary over a 16-month period, 45 were discharged by the surgeon in charge, in the belief they were cured, and 15 died. During this 16-month period no one was re-admitted.

Admission of leprosy patients to the Native Infirmary was voluntary and, according to the place of origin given by those who were in-patients from 1812–13, many residents had travelled considerable distances for treatment. Many who lived by begging were inhabitants of other places where they had families, though some had relatives in Madras or its immediate neighbourhood. The city of Madras was a magnet for leprosy sufferers in the presidency. The superintendent of the 1911 Madras Census reported that tabulation of each leprosy sufferer's birthplace showed that there was little or no migration among leprosy sufferers except into Madras City.[23] The presence of the Monegar Choultry and Native Infirmary and, later, the Madras Leper Hospital, together with the greater opportunities in a large town for gaining alms, encouraged immigration into Madras. The largest number who came to the infirmary were from Tamil-speaking areas with others from Malayalam-, Telugu- and Kanada-speaking regions. In some cases, leprosy sufferers travelled together, possibly as members of the same family. Three recorded in the Native Infirmary's records as 'caffres' from Goa, aged 50, 28 and 20 were admitted together on 13 January 1813, and discharged together on 2 May.

As suggested by the numbers of leprosy sufferers who remained in employment in Madras and Calicut, where hospital care was easily available, leprosy sufferers tended to seek hospitalization only when the disease became very severe.[24] Some, however, did not wait until illness and poverty forced them into a hospital, but planned their admission in advance. In 1891, Ponnambala Pillay, a poor leprosy sufferer of Mayavaram, advised the district magistrate at Tanjore that since he was unable to take care of himself and had no relatives or friends to take care of him, and there was no ac-commodation for leprosy sufferers in Mayavaram, he would like to be admitted to the leprosy hospital at Madras. In response, the district magistrate agreed to pay the costs of Ponnambala Pillay's transport by rail to the Madras Leper Hospital.[25]

The nature of the institution

While there were links between the confinement of leprosy sufferers and of prisoners, it is misleading to understand the leprosy hospital as a centre of discipline and power. The hospitals of the Madras presidency were primarily medical institutions of voluntary residence. Until the Government of India's 1898 Lepers Act there was no legal provision for forcible confinement of leprosy sufferers in India, and even when passed, the Act was not applied in the Madras presidency during the nineteenth century.

Even so, the British medical officers in charge of the Madras Leper Hospital considered that a degree of patient control was necessary for the effective management of the institution and for the benefit of the patients. In discussions over whether or not the apothecary should reside on the premises of the leper hospital, Surgeon-General E.G. Balfour stressed the need for the apothecary to be permanently on site since he was in effect the superintendent of the institution. As such, Balfour argued, the apothecary was required to live close to the inmates 'over whom a constant though careful and gentle control is needed'. However, the autonomy of the patients and their freedom to come and go as they pleased was also acknowledged. Surgeon van-Someren, commenting on the Madras Leper Hospital in 1860, lamented that many who became in-patients, 'weary of the monotony, rebel against the restraints and discipline of the hospital, and seek their discharge after periods of various duration', only to return once their disease became aggravated.[26] The leprosy hospital was not a 'total institution' in Goffman's terms. Although, to some extent, the sufferers did 'together lead an enclosed, formally administered round of life', control was not complete. Further, technically at least, the leprosy sufferers were not prisoners, cut off from society, but were free to leave.[27]

Design

The institutions designated for the care of leprosy sufferers in the Madras presidency bore little resemblance to Bentham's Panopticon. Hospitals were designed to provide a healthy and comfortable environment for patients which as far as possible complied with their cultural and religious requirements. Although plans for the first Madras Leper Hospital have been lost, there are physical descriptions of the first permanent buildings provided to house leprosy sufferers. There was little difference between the first wards of the Native Infirmary built in 1799, which consisted of two large blocks each of two wards, surrounded by verandahs, and the bungalow built to house leprosy patients in 1816 on four-and-a-quarter grounds (a ground equals 2400 sq ft) of land opposite the Monegar Choultry. The emphasis in building design was on the patient's health. The huts adjoining the infirmary were removed to preserve the circulation of air to the building and, on the advice of the medical officers, a verandah was added.[28] Proper ventilation was also of primary concern in the extensions to the hospital in 1840. Surgeon Lawder argued that while the wards for leprosy patients, 'followed the same cheap style of buildings as that which is always adopted for native hospitals and other ordinary purposes', particular attention to ventilation was needed in designing the new hospital buildings since they were intended for 'accommodation of persons labouring under so loathesome a disease as leprosy, which from its nature does and must always render the surrounding atmosphere impure'.[29] By the 1840s, the British had perceived a clear link between poor ventilation and the spread of disease. The concept was later detailed in the *Madras Manual of Hygiene* edited by King, who urged the necessity of hospital ventilation since the air in such institutions was 'liable to pollution by morbid effluvia from sick and wounded, from excretions, dressings, poultices and bandages'.[30]

While there was no ring of buildings under the fixed gaze of a central tower 'ensuring the automatic functioning of power',[31] the Madras Leper Hospital was enclosed by a high brick wall. The wall was built on the advice of the medical officers in charge of the Monegar Choultry and Native Infirmary who argued that such enclosure was 'necessary in a disease of the kind, these patients are afflicted with, to keep them from intercourse with others and at the same time to guard against any improper food being admitted'.[32] The British medical officers perceived fear of close contact with leprosy sufferers by Indian patients and were themselves uncertain about the disease's mode of transmission. Further, they were concerned that, for therapeutic reasons, the patients' diet remained closely controlled. While the management committee of the Monegar Choultry advised the governor that leprosy was still at that time 'deemed by medical men not to be infectious', the committee believed it necessary to ensure that leprosy sufferers be prevented from contact with others,[33] presumably to prevent transfer of the disease. Such a wall was not a typical architectural feature of medical hospitals. However, lunatic asylums in India were often enclosed,[34] as were jails, suggesting that leprosy was not regarded solely as a medical condition. The presence of the wall was an intimation of incarceration.

Caste, race and gender

By the close of the nineteenth century, the Madras Leper Hospital was a large complex segregated on race and gender lines. There were three wards (with a covered way) totalling 24 beds for European men, three wards totalling 72 beds for Indian men, one ward of 28 beds for Indian women and one of 16 beds for European women. Boys and girls were accommodated in the male and female wards respectively. The Leper Hospital was notable for its lack of caste wards. The Government Ophthalmic Hospital, which, like the Leper Hospital, served both the European and Indian population, had by the close of the nineteenth century one ward for caste females and one for caste male patients.[35] The lack of caste wards in the Leper Hospital was not necessarily a consequence of the high number of low- and outcaste patients, and the paucity of high-caste patients admitted. The Native Infirmary, completed in 1799, included separate facilities for brahmins. The Infirmary was built to accommodate the Indian poor of every caste and a separate ward was provided for brahmins in the belief that they would otherwise be reluctant to attend.[36]

It is more likely that the absence of caste wards reflected the reluctance of the Madras government to spend more than absolutely necessary on leprosy care. The Madras government's decision to house Eurasian leprosy sufferers with European ones, despite strong lobbying from the Friend in Need Society for separate accommodation, was based on the view that the number of Eurasian patients was too small to warrant expenditure on additional accommodation.[37] Even so, the Madras Leper Hospital was among the first to provide separate accommodation for European and Indian patients, with Eurasians included with Europeans. In Calcutta, such accommodation was proposed only in 1890.[38] While the lack of separate accommodation could well have been a deterrent to the admission of brahmin leprosy sufferers into hospital, their absence also meant there was no pressure on the Madras government to provide for them. The issue of loss of caste status for brahmins who entered the hospital was to become important in the framing of the Lepers Act.

Hospital life

While caste concerns were not adequately catered for in leprosy asylums, some effort was made to provide facilities for the spiritual care of leprosy sufferers. In 1886 the majority of patients at Palliport Lazaretto were Roman Catholic, reflecting the Portuguese history of the region, with only one Protestant, a convert from Roman Catholicism. One ward in the women's section was employed as a

Roman Catholic chapel in which the patients themselves officiated. In the Madras Leper Hospital there were two small chapels, one Church of England and one Roman Catholic, in which clergymen of each denomination held services. In addition, a place was set apart for Hindus to worship.[39]

Considerable efforts were made at some leprosy hospitals to reduce the tedium of hospital life and to encourage leprosy sufferers to remain by reconciling them to its restrictions. At the Madras Leper Hospital, patients were encouraged to employ themselves in the cultivation of fruit and vegetables and, as an added incentive to work, the patients themselves kept the proceeds of the sale of the garden produce. Gardening was a congenial activity for both Indian and Eurasian residents but was especially appreciated by the Eurasian patients. At Palliport Lazaretto, Day ascribed the difficulty of managing patients to their lack of occupation and, in 1860, proposed allocating plots of land for cultivation to those who wished to engage in some activity. He hoped that by giving the leprosy sufferers a diversion and sense of purpose they would become more amenable to care.[40]

In comparison with the ordinary poor and sick housed at the Madras Infirmary and Native Poor Asylum in 1812–13, those with leprosy received a generous dietary and living allowance. Leprosy sufferers in the asylum received per head per month approximately 1.9 marcáls of rice, 10 seers of mutton and 1 vis of ghee,[41] bread, occasionally one chicken, 'curry stuff' and beetlenut cash. The luxury of receiving beetlenut and tobacco which, according to the Native Sick Medical Code (xv:v), was permitted only to the 'native sick' in the Lunatic Asylum and Eye Infirmary, was extended to the Madras Leper Hospital in 1840. In addition to food, they received firewood and bratties (cow dung fuel cakes), cloth, red caps, worn as part of the asylum clothing, and lamp oil as required.[42] It is likely that the red caps were not simply for protection but signified that the leprosy sufferers were an asylum population. Similar red caps were worn by 'lifers' in the Madras presidency jails, certainly at the close of the nineteenth century and very likely also in the earlier decades.[43] The meaning of such caps highlighted the ambiguity of the leprosy sufferers' status, whether they were prisoners, or patients who were free to leave the hospital.

While patients were free to leave the asylum and considerable effort was made to cater for their needs, there was, however, some measure of discipline applied within the hospital. According to the

Standing Orders of the Government Leper Hospital, Madras, approved in 1895, several actions warranted discharge. Diet was an area of particular control, with patients permitted only hospital food and drink. Except with the special permission of the superintendent, patients were forbidden to give away any portion of their diet or to remove any article of diet from the hospital. Patients who violated these rules were discharged. Giving any gratuity, food or drink to hospital servants was also grounds for discharge.

Further, any visitors who brought food or drink of any kind into the hospital, or removed or ate any part of the patient's diet were to be handed over to the police. This rule was not unique to the Madras Leper Hospital but applied also to the General Hospital. It was disputed in 1877 by the medical branch of the Home Department as being excessive and of doubtful legality since it enabled a person to be dealt with as a criminal for the sole offence of breaking hospital rules. Even so, when queried by the Home government, the rule was upheld by the Madras government as appropriate to the General Hospital's effective management.[44] The imposition of power within hospitals, while firm, was at least open to discussion.

Despite the severity of the Madras government's ruling on food, generally within asylums there was minimal restraint on leprosy sufferers. In the Madras Leper Hospital, smoking in the wards was permitted only to those who were bedridden, otherwise it was allowed only on the verandahs. Quiet was expected in the wards after 9.00 p.m. The principal restrictions applied to visitors. Visitors were permitted only between 3.00 and 6.00 p.m. and patients were not allowed to converse through the railings surrounding the Criminal Leper Ward. However, since leprosy patients were permitted to leave the asylum, in reality there was little limit placed on the relationships they formed. Further, as H.D. Cook, surgeon in charge, wrote of the Madras Leper Hospital in 1889: 'there are no restrictions for a husband to visit his stricken wife, or *vice versâ*, and births are not uncommon in asylums set apart for lepers'. Absence from the hospital without obtaining leave from the superintendent was, however, grounds for discharge and there was some restriction on readmission for discharged patients. Those discharged from the hospital for misconduct or who left without permission could not be readmitted except on approval of the superintendent. However, in cases of emergency such a patient would be readmitted and, if possible, detained at the hospital to ensure they remained for sufficient time to recover.[45]

For the leprosy sufferer there was scope for resistance to hospital authority and for the negotiation of power. In the Madras Leper Hospital, resistance by leprosy sufferers was indeed, to use Ignatieff's terms, 'a constitutive element of the history of power'.[46] Diet, being one of the few aspects of institutional life worth looking forward to, was often a matter of negotiation between patients and hospital staff. In 1873, patients in the Madras Leper Hospital, dissatisfied with the diet regimen introduced in 1865, bartered away food items in contravention of hospital regulations. Although giving away food typically led to a patient's discharge, the hospital authorities, after stopping the practice, listened to the patients' grievances. The deputy inspector-general, who favoured the existing diet for its benefits to the patients' health, 'told the inmates his views in his usual gentle manner' but ultimately yielded to the patients' request. A revised diet, securing the provision of curry and rice to the patients, was decided.[47] However, the quality of diet did not cease to be an issue. A new diet was trialled in the Madras Leper Hospital in 1874 and Indian patients complained that, although eating the same breakfast each day was acceptable, eating the same evening meal was intolerably dull. In response, van-Someren, the surgeon in charge, recommended a rotation of evening meals for a small increase in price which 'would make the residents much happier' as well as assisting the efficacy of medical treatments.[48]

In some instances where leprosy sufferers found hospital authorities to be less co-operative with their requirements, they appealed for outside help against the hospital staff. In 1873, van-Someren proposed putting some leprosy patients at the Madras Leper Hospital to light work, enabling the hospital to reduce its staff by two coolies, one female toty (labourer, messenger) and a tailor, thus saving considerably on wages. Reflecting Victorian concepts of morality, van-Someren also suggested that such employment would be beneficial, both physically and morally, to the leprosy patients as 'they would no longer be eating the "bread of idleness"'. While the British distinguished the leprosy sufferer who was vagrant because of illness from the wider vagrant class who were perceived as poor due to the vice of indolence, it was believed that leprosy sufferers could also be tainted with laziness. Of the 68 patients requested to perform various kinds of light work, 43 refused to comply. Given an ultimatum to comply or be reduced to a spoon diet consisting of broths, gruels and, for Indian patients, conjee (rice porridge), the patients instead went in a group to the Town Police Office to place their case. The magistrate, however, being informed of the case from the hospital's perspective, directed the patients to return to the hospital and to comply with its requirements, which they did.[49] There was no apparent distinction made between Indian, European and Eurasian leprosy sufferers in making the request to work. The only criterion was the health of the patient.

Evidently, by the late nineteenth century the leprosy sufferers at the Madras Leper Hospital were not only aware of their freedom to leave the asylum and willing to exercise that right but were confident that British legal authorities would listen to their case and possibly act on their behalf, even though the magistrate ultimately decided against them. Further, the fact that they were more prepared to submit to legal authority than to the institution's implies that they did not consider institutional discipline to be in any way binding. Expulsion from the asylum, the ultimate discipline, was only an effective deterrent to non-compliance if the patient did not wish to leave. At the Palliport Lazaretto, the same patient was discharged for 'heavy drinking', readmitted three months later and discharged again within the month for theft.[50]

Patient or prisoner? The ambiguity of the leprosy sufferer's status

Until the 1898 Lepers Act, a leprosy hospital was legally a place of voluntary residence with relatively little regulation of in-patients. However, there was some ambiguity regarding the leprosy sufferer's status in the asylum, specifically whether the leprosy sufferer within the institution was a prisoner or a patient. In British discourse on leprosy in Madras, the leprosy sufferer's primary status was as a patient and an object of charity. The notion of compulsory confinement, however, began to be debated as soon as separate hospital accommodation for leprosy sufferers became available at the Monegar Choultry and did not remain for long a concept primarily motivated by the best interests of the sufferer's health. The British desire for greater control over leprosy patients, particularly their isolation from both the Indian and British residents in Madras, was not only due to a concern for the leprosy sufferer's health and hopes to control the spread of the disease. It was also a means of contributing to the maintenance of social distinctions by preventing continual encounters between British residents and leprous beggars.

A leprosy sufferer who was also poor and vagrant and who lived by begging was far more susceptible to confinement than someone with leprosy who lived at home or lived by other forms of employment. Once the permanent asylum for leprosy sufferers had been completed, the management committee of the Monegar Choultry and Native Infirmary began to explore the possibility of using police authority to confine leprosy sufferers who earned their living by begging on the streets. The proposal to use legal force to remove these people from the street and into the asylum marked a sharpening of the distinction between the leprosy sufferer who sought medical attention and care, and the one who lived by begging and, as such, posed a nuisance to the public.

In 1817, during discussions concerning the revision of the Madras Police Regulations, the management committee recommended to the Madras government that the Monegar Choultry extend its role in clearing the streets of pauper vagrants to include specifically those suffering leprosy.[51] Under the Madras Police Regulations, the police had the power 'to commit to labour on the public roads, persons placing themselves in the highways together alms' [sic]. While noting that 'a leper is not a fit object of such punishment', the management committee of the Monegar Choultry and Native Infirmary nevertheless hoped to enlist the authority of the police and magistrates in order to bring leprosy sufferers into the asylum. The management committee was concerned both to extend the 'benefits of the institution' to leprosy sufferers who had no other means of support than begging and to benefit the Madras public by ensuring 'an early removal of the loathsome beings who are now but too frequently seen in our roads and highways'. Leprosy sufferers on the street were perceived not solely as deserving of charitable and medical aid but as 'wretched beggars', an eyesore to be removed from the public, particularly the British public, gaze.[52]

Since police were empowered only to put vagrant paupers to work on the roads, not to apprehend a leprosy sufferer, the management committee suggested to the governor, Hugh Elliot, that the police regulations be altered to enable police to send to hospital leprosy sufferers 'exposing themselves to gather alms on the highways'. The committee requested that leprosy sufferers, prior to their being sent to the asylum, be examined before a magistrate to ascertain whether they could be supported by friends or family, thereby reducing the number depending on the choultry's resources. It was only those with no means of subsistence other than public begging on the street whom the committee recommended that the police pick up and bring to the leper hospital for confinement. Given that the disease was believed to be incurable, the committee considered that 'confinement . . . of a person in this state, who has no other means of subsistence but the charity of others may be considered perpetual', thus entailing a substantial demand on the choultry's financial resources. Throughout the nineteenth century, at local, presidency and Government of India levels of authority, such concerns to economize continued to place firm limits on any practical efforts to confine leprosy sufferers.

In cases where the leprosy sufferers had friends or family, the committee recommended that the leprosy sufferer be placed in their charge, the friends or family 'undertaking under a penalty that he shall not again be found exposing himself on the highways'. Where the leprosy sufferer's family was resident outside Madras, the committee suggested 'he be conveyed to his home in police custody, and delivered to the head of his village, or other police officer on the spot, with a charge that he be not allowed to quit it'. However, if leprosy sufferers had means of support through friends or family but preferred residence in the asylum, they were to be admitted.[53]

Only those leprosy sufferers who were vagrant and without family and friends to support them were to be the subject of forced confinement, the same group of who were to be the target of the Lepers Act. Constraints were to be placed on the leprosy sufferer by family and friends. However, sufferers who had such alternative means of support were perceived by the committee as having options, being free to choose whether or not to receive care in the asylum. The Committee for the Police of Madras confirmed the tenor of the Monegar Choultry and Native Infirmary's Committee's recommendations:

> With regard to Lepers we think, if they have other means of subsistence of their own, they will not expose themselves in the streets to beg – if they are beggars, the magistrates must take them up, and in their refusing to retire to their own houses, must deal with them as with other objects of disgust, who beg in the streets – the Magistrates need not work them on the roads, it would be obviously improper to work them, or any other sick prisoners; they must be kept in a proper place, separate from other prisoners – the choultry may be the best place.

Police contempt for vagrant beggars and leprosy sufferers was clear from the Police Committee's Report. Beggars were 'objects of disgust' and beggars with leprosy were represented as vermin, 'infesting the streets'.[54]

Ironically, while enclosing the accommodation for leprosy sufferers with a wall and advocating the removal of leprosy sufferers from the streets by the police, the management committee was concerned that the Leper Hospital would be seen by leprosy sufferers as a 'place of terror' rather than a 'charitable and benevolent institution' for their care. The committee hoped that the selection for placement in the leprosy hospital of only those without family or friends would not deter others from seeking admission simply for medical treatment.[55]

Such distinctions were lost, however, on leprosy sufferers struck by the ominous presence of the hospital wall. Many leprosy sufferers were not entirely convinced of their status as patients in the Madras Leper Hospital and feared that once they had entered the asylum they would be prevented from leaving. Prevention of the spread of leprosy and protection of the leprosy sufferers' health by ensuring the correct diet provided the British rationale for building the wall. However, for the leprosy sufferer it was the first tangible evidence of British moves towards their confinement. Rumours had already gone abroad that the hospital was indeed a prison for incarceration of the leprosy sufferers in Madras. As van-Someren, surgeon in charge of the Madras Leper Hospital, commented in 1860, the use of 'high dead walls' to enclose the buildings of the Madras Leper Hospital, 'by giving the hospital the aspect of a prison, serves to deter many from seeking asylum within its precincts'.[56]

Although it is not clear whether or not the changes in the police regulations recommended by the Monegar Choultry's management committee were implemented, it is evident that, by 1817, the committee made a clear distinction between the homeless vagrant and the voluntary in-patient. Only the vagrant leprosy sufferer was susceptible to forced perpetual confinement. For those who had some other form of socio-economic support, the Leper Hospital was to remain a place of voluntary treatment. To the rhetoric of charity had been added mechanisms for the compulsory confinement of the destitute and homeless vagrant with leprosy which allowed those with family or other financial support to choose whether or not to enter the asylum. Rhetorically at least, the Madras Leper Hospital had become a prison for the homeless vagrant with leprosy, while for the leprosy sufferer with family support it remained a hospital.

The leprosy sufferer with fewest socio-economic supports was most vulnerable to compulsory confinement and, once within hospital, was most susceptible to being regarded as a prisoner.

Despite the shift in rhetoric, the leprosy sufferer remained, in practice, primarily an object of charity and, as such stretched the limits of accommodation at the Madras Leper Hospital, finally forcing its permanent expansion in 1840. Overcrowding in the asylum was partly a consequence of the disjunction between the leprosy sufferers' medical needs and the police responsibility to clear the streets of vagrants. The management committee of the Monegar Choultry reported to the governor-in-council in 1823 that leprosy sufferers discharged from the Native Infirmary on the grounds of requiring no further medical treatment were then apprehended and sent to the choultry by the superintendent of police as vagrants needing shelter. Since it was impractical to assign them a separate ward in that section of the institution, they were then, 'as helpless objects of charity re-admitted into the leper hospital'. The committee argued for the expansion of the Leper Hospital on the basis that the leprosy sufferer was an 'object of charity' and claimed it was the 'indispensable duty' of the institution 'to provide accommodation for persons of the above description, who are generally destitute of friends and means of obtaining their livelihood'.[57]

The extension of the Madras Leper Hospital in 1840 revived the issue of the forced confinement of leprosy sufferers. The superintendent of police welcomed the opportunity promised by the additional accommodation to apply specifically to leprosy sufferers the Madras government policy of removing vagrants from the streets as recommended in 1817 by the Monegar Choultry's Committee. He looked forward to being able 'to clear the town of Madras of all lepers . . . who have not the means of supporting themselves'.[58] Further, the expansion of the leper hospital focused attention again on the wall surrounding the hospital, which can be seen as a metaphor for the problematic role played by the hospital in the confinement of leprosy sufferers.

By 1840, Leper Hospital and Madras government authorities agreed that a wall enclosing the hospital was necessary. However, the height of the wall continued to cause considerable debate. The military board, to which all plans for building were submitted, estimated the cost of building two new wards with the requisite outhouses and a seven-foot-high compound wall adjoining the existing accommodation for leprosy sufferers at Rs 8268-4-3. The military board's estimate sparked discussion on the appropriate height of the wall. The principal arguments in favour of an increase in the wall's height were medical. Surgeon Lawder, in charge of the Leper Hospital, considered that a wall seven feet high was 'wholly insufficient'. He argued strongly on the same grounds that justified the original decision to enclose the male leprosy ward in 1819, that an increase in height to 10 feet was an 'indispensable necessity ... in order to prevent communication between the patients and people outside and the sale of their clothes and consequent spread of the disease'. He continued that if the wall were not raised, 'the object of the additional wards will be in a great measure defeated'.[59]

On 1 October 1839, the Madras government referred the matter of the wall's height to the medical board for report, noting their concern that a 10-foot-high wall might prevent the free circulation of air. The medical board confirmed the government's concern, but accepted Lawder's contention that 'the height of seven feet is insufficient to prevent communication between the leper patients, and the people outside the hospital'. Rather, the board recommended that the wall be raised to the height of $8\frac{1}{2}$ feet 'which without materially obstructing the free circulation of air, would effectually cut off all intercourse between the inmates of the Hospital and other persons'. The government accepted the opinion of the medical board and on 29 October 1839, authorized the wall to be raised from 7 to $8\frac{1}{2}$ feet and requested a revised estimate of costs.[60]

The discussions over the wall embodied the tensions inherent in British efforts to confine leprosy sufferers in Madras. By the 1840s there was agreement between different levels of medical authority and the Madras government that a wall was necessary to prevent the spread of the disease. However, the final height of the wall represented a compromise between disease prevention, the leprosy sufferer's own needs and the Madras government's desire to economize. To some degree there was agreement among those in power on the perceived need to enclose the vagrant population. However, the degree of unity was at best modest and represented a compromise between economics and competing understandings of the purpose of confinement.

The links between leprosy sufferer and prisoner continued to develop despite the arguments in favour of the confinement and segregation of leprosy sufferers being framed primarily in medical terms. On the recommendation of the medical board the commissariat department was to provide food for Indian in-patients in the new Leper

Hospital in the same way as they provided for prisoners in the presidency jails.[61] This decision did not necessarily represent a deliberate racial policy of linking Indian leprosy sufferers with prisoners. Measures for the confinement of vagrant leprosy sufferers included all races. The link between the jail system and Indian patients was more a reflection of the logistical pragmatism of providing food appropriate to Indian tastes on a large scale since vagrant Indian leprosy sufferers formed a substantial majority in the leper hospital.

In December of 1840 the ambiguity of the leprosy sufferer's status as patient or prisoner was made abundantly clear. On 3 December, Lawder reported the departure without permission of four leprosy sufferers over the wall of the Madras Leper Hospital as an 'escape'. Fearing that 'the same thing will occur frequently' he requested Madras government approval for raising the wall 'as soon as possible'.[62] This was not the first time that patients had 'escaped'. Eleven male patients had climbed the hospital wall on 15 July of the same year. Clearly the surgeon in charge of the Leper Hospital considered unauthorized departure from the hospital to be 'escape', implying that confinement was a significant role of the hospital.[63]

Lawder had recommended to the superintending surgeon that the wall be raised after the July escapes. However, although the escape was reported to the police with the intention that the patients would be caught and returned to the hospital, no action was taken in raising the wall. That some of those who had left in December had done so earlier in July indicates both that the police had succeeded in returning them to the hospital and that they had genuinely wished to leave. On each occasion the patients took their hospital clothing and the blankets supplied to them but nothing more. There appeared to be no deliberate intention of theft, but rather to leave with the basic necessities of clothing and protection from the weather. Lawder was concerned both to prevent the escape of patients and to ensure that hospital property would not be removed. He argued that since such departures had occurred twice it was necessary to raise the walls surrounding the yards of the male ward of the hospital from $8\frac{1}{2}$ feet to 12 feet to prevent their more frequent occurrence. Despite the prevalence of arguments in favour of confinement to reduce the spread of leprosy, Lawder's demand to raise the wall implies that it existed also to ensure the hospital's security as a place of permanent detention for leprosy sufferers. That the patients considered it necessary to make a physical escape over an $8\frac{1}{2}$-foot wall, a difficult task even for someone in good health, indicates that they believed that their residence in hospital was a form of imprisonment and that they were not free to leave.

Indeed, this was precisely Lawder's view. In his report for the second half of 1840 published by the later surgeon at the Madras Hospital, van-Someren, Lawder outlined his policy of admission to the Leper Hospital. Since many leprosy sufferers in the hospital were not voluntary patients but were taken from the streets by the police 'in a stinking state, consequent on want' and sent to the Leper Hospital. Lawder explained: 'I considered it my duty to retain them as well from motives of humanity, as from a wish to seclude them from public view, and also to check the extension of the disease, which has for some years become very apparent at Madras.'[64]

In response to Lawder's demand for raising the height of the wall, G. Pearse, secretary to the medical board, asked the Madras government to clarify whether the leprosy sufferer's status was more correctly that of a patient or a prisoner. Given that Lawder considered raising the height of the wall to be the only means of preventing the leprosy sufferer's escape, the board requested to be advised

> the wishes of Government respecting the permanent detention of those persons in the institution, contrary to their wishes; or whether, like patients in other hospitals, they are to be allowed their discharge on application, as in the latter case, it will be unnecessary to incur the expense of heightening the wall, or to adopt further measures for their security.[65]

The Madras government clearly decided in favour of the leprosy sufferer as a patient rather than a prisoner, resolving, on 22 December 1840, that it was unnecessary to raise the wall surrounding the leper hospital since 'it is not the desire of the Government that lepers should be detained in the Monegar Choultry against their inclination'. That the residents escaped in hospital clothing was reported to the police, but was considered by the government to be a matter of theft,[66] an issue entirely separate from the forced confinement of leprosy sufferers. From the medical board and Madras government's point of view, leprosy sufferers were not prisoners, despite the opinion of the surgeon in charge and the involvement of the police in bringing them to the asylum.

The position of the Madras government was consistent with that of the British government which was firmly grounded in the desire to contain the disease not the sufferer. Following the Royal College of Physicians' finding that leprosy was not contagious, the secretary of state for the colonies strongly expressed his abhorrence for the confinement of leprosy sufferers. He declared in 1863: 'that any laws affecting the personal liberty of lepers ought to be repealed', and that any form of restraint 'merely authorized and not enforced by law, ought to cease'. As a consequence, leprosy hospitals were closed and segregation measures ended in Dutch Guiana and many British colonies, including British Guiana, St Kitts, the Molucca and Oeliasar Islands, and in Java.[67]

In compliance with the Madras government's resolution, Lawder agreed no longer to endeavour to retain against their will patients sent by the police to the hospital. However, he accepted the ruling with regret, writing in his report:

> I feel that the lives of many brought to the hospital by the Police under such circumstances are saved, who will now fall a sacrifice to their own obstinacy in declining to remain in hospital, particularly as I generally find the worst, and most disgusting cases to behold, are those who are most anxious to remain at large.[68]

In the Madras presidency during the first half of the nineteenth century, confinement of the leprosy sufferer who lived on the margins of society as a beggar or vagrant was not endorsed by all levels of British authority and certainly not by the leprosy sufferers themselves. In 1840 there was a clear difference of opinion between Lawder and the Madras government over the Leper Hospital's role in confinement. The escape of leprosy sufferers from the hospital, the ultimate resistance, called into question Lawder's imposition of confinement. Rather than endorsing the practice of confinement by the hospital's medical authority, the Madras government intervened to allow the leprosy sufferers greater freedom.

The blurring of the lines between the leprosy sufferer as prisoner or patient in the Madras Leper Hospital reflected the complexity and incoherence of British notions of confinement and charitable care for leprosy sufferers in Madras. Confinement of leprosy sufferers was not a coherent expression of British power. Rather, it was a consequence of the dual pressures from both government and medical officers to provide medical care and shelter for leprosy sufferers and to clear beggars from the streets of Madras. It was not the Madras government which provided the initiative for the forced confinement of leprosy sufferers. Medical officers, while emphasizing their role in the charitable and medical care of leprosy sufferers, advocated the selective detention in hospital of those who had no other means of subsistence than begging, thus creating two categories of hospital resident, the voluntary patient and the involuntary prisoner. Any initiatives for confinement were, however, curtailed by the economic prerogative of colonialism, as were efforts to impose British authority through the use of British medicine. Together with concerns of charity and the desire to remove vagrants, financial restrictions on both the Leper Hospital and the Madras government contributed to the Monegar Choultry's advocacy of selecting for confinement only those without other socio-economic support. The Madras government's reluctance both to extend institutional accommodation for leprosy sufferers and to build, then to increase the height of, the wall surrounding the Leper Hospital, further slowed the development of institutional structures for confining leprosy sufferers.

Despite the ambiguity of the leprosy sufferer's status, the Madras Leper Hospital remained principally an institution for the care and treatment of leprosy sufferers rather than for their forced confinement, a status which came into question again with the discussion of the Lepers Act at the end of the nineteenth century. Further, negotiations over diet and labour in the Madras Leper Hospital and the complex response to British medical treatment of leprosy, indicate that the conditions of confinement within the asylum were almost always open to dispute and adjustment. Disease did not engender passivity: by resistance to institutional confinement and British medical treatment, the leprosy sufferer challenged and disrupted the imposition of British colonial authority.

3
Colonial Medicine in the Indigenous Context

During the first part of the nineteenth century, British leprosy treatment in the Madras presidency was guided principally by the inclination and judgement of the medical officers directly involved in the care of leprosy sufferers. There was some monitoring by the medical board and presidency government, but little interference. It was not until the 1860s that the attention of the Home government and Government of India was drawn to leprosy treatment, and the more distant levels of government began to direct British treatment of leprosy at the local level.

The relationship between British and indigenous medical practice changed considerably during the nineteenth century, and the treatment of leprosy, particularly the Indian patient's response to treatment, reflected the tensions and accommodations developing between the two systems.

The indigenous medical context

By the early nineteenth century, political and social upheaval, caused in part by the mughal incursions, had brought about a decline in the practice of indigenous medicine in India.[1] Even so, when British medical officers in Madras became involved in the medical treatment of leprosy in the early nineteenth century, they did not step into a treatment vacuum. There was already a complex indigenous medical response to leprosy which influenced British attempts to treat the disease. It is difficult to describe precisely the indigenous tradition which the British medical officers encountered in the Madras presidency. There was no fixed, pure tradition of indigenous medical practice in south India, and the principal southern indigenous textual tradition, the Hindu tradition of *citta vaittiyam*, or Siddha medicine, is notoriously elusive. Though often regarded as a Tamil variant of Ayurveda, the Sanskrit medical tradition, many practitioners of Siddha medicine claim that it arose independently of Ayurveda. siddha medicine was believed to have originated with the 18 siddhas, poets and philosophers of south India, renowned for their piety and wisdom. The siddha was one who had gained *citti*: 'power, prowess, strength, ability', and had advanced to gain eight kinds of supernatural power including the capacity to shrink to the size of an atom, and dominion over the animate and inanimate worlds. The Tamil siddhas were part of a wider, rather esoteric but popular 'medieval' Indian agamic, tantric and yogic tradition which ran parallel to the *bhakti* movement. Like followers of *bhakti*, the siddha criticized the caste system and brahmanic superiority but they opposed the more emotional aspects of *bhakti* devotion. It is unknown when Siddha began in south India or the exact number and identity of the siddhas, though traditionally there were 18 siddhas of various nationalities and castes, including only a few brahmins.[2]

The siddhas traced their origin to Agastyar and the writings on mysticism, alchemy and medicine ascribed to him. Although the historicity of Agastyar has been the subject of considerable debate, it is clear that in addition to the legendary figure of this brahmin philosopher and poet there were several siddhas who bore the same name and composed a number of acclaimed works on poetry, alchemy and medicine. Among the essentially esoteric Siddha works some of those ascribed to Agastyar were notable for the effort to make accessible to the reader the aetiology, symptoms and appropriate *materia medica* for treatment of disease.[3] The greatest of the siddhas wrote between the tenth and fifteenth centuries, with the exception of the renowned Tirumūlar who according to legend lived in about 3000 BC, but who probably walked the earth during the seventh century. The siddhas did not develop a unified, coherent system of philosophy, although most refuted karma and reincarnation, believing that eternal life and youth could be attained by attention to both body and soul in this life, particularly through medicine and yogic meditation. Even so, the notion of a karmic cause of disease was present in Siddha medicine, most likely because the substance of all myths concerning the sage Agastyar was Sanskritic in origin and character.[4]

Whatever the origins of Siddha medicine, a strong correlation did exist between the conceptualization of disease and methods of cure in the Ayurvedic and Siddha traditions. According to a text attributed to Agasthyar, 'a good physician must be thoroughly skilled in the commentaries on the *Ayurveda*'.[5] Both systems were based on the idea that medicines, the human body and the universe existed in close relationship: whatever exists in the universe exists also in the body. Whitelaw Ainslie, a Madras-based medical officer and author of the first British south Indian *materia medica*, suggested a further distinction between the Sanskrit and Tamil medical traditions. Ainslie noted in the *Materia Indica*, published in 1826, that the Tamil medical texts ascribed less to magic and mythology in their medical lore than the Sanskrit medical texts, and attended more to detailing symptoms of disease and describing their precise treatment. Even so, he hastened to add that the Tamil texts were not entirely bereft of references to the supernatural. Ainslie observed that Tamil texts like the Ayurvedic, 'evince a firm conviction in the belief and intervention of evil spirits, and offer many curious rules for averting their machinations'.[6]

Tamil medicine shared with Ayurveda the notion that the harmonious action of the elements, five in Siddha and seven in Ayurveda, and the balance of the three humours, air, bile and phlegm, constituted health and life. Disease was thought to be caused by the derangement of humours, and cure was sought by the restoration of their proper balance through a strict treatment regimen, prohibition of certain foods and the prescription of appropriate medicines. Diet was an important aspect of treatment on the basis that all foods were of either a heating or cooling nature. Diagnosis of disease in both Ayurvedic and Siddha medicine was made by attention to body temperature, the appearance of the eyes, mode of speech, colour of the face and body, condition of urine and stools, the state of the tongue and the reading of the pulse. In Siddha medicine the reading of the pulse drew on yogic powers of concentration, and not only diagnosis but prognosis of disease could be discovered. Diagnosis of urine was also specialized and included reading the surface tension and patterns formed by adding a drop of oil.[7]

While sharing with Ayurveda a basic concern with curing disease, Siddha medicine, unlike Ayurveda, which assumed reincarnation to be the path to immortality, was motivated by a desire to achieve bodily immortality in this life. This aim was expressed philosophically and medically in the siddha's attention to alchemical transmutation. Although alchemical ideas did exist in the Sanskrit texts, Siddha, as a consequence of the fundamental philosophical difference between the two systems, placed far greater emphasis on the use of imperishable inorganic materials in medical treatments than did Ayuveda, and gave alchemy a more central and soteriological significance. In particular, Tamil Siddha texts refer to *muppū*, meaning a mixture or compound of three salts, a substance, like the European philosopher's stone, believed able to transmute base metals to gold and to rejuvenate the human body. The Sanskrit textual tradition does not refer to such a substance. This emphasis in Siddha on inorganic treatments is held by some to be the basis for the Siddha claims for an origin independent of Ayurveda.[8]

During the nineteenth century, Tamil medical practitioners drew substantially on medical works ascribed to Agastyar, which were the most highly regarded of those available. Tamil medical texts tended to follow a particular pattern. Each was prefaced with a topographical memoir describing the local climate and soil conditions, followed by a *materia medica* including a description of the correct seasons for gathering vegetable drugs, and a description of the modes of preparation and dosage of medicaments. The characteristic patterns of the pulse were described in detail and rules were given for the diagnosis and prognosis of disease, with instructions for diet and exercise.

The southern medical textual tradition was not, however, exclusively Tamil, and included the Ayurvedic Sanskritic texts of Caraka and Suśruta which were available in Malayalam, Canarese [Kanada] and partially in Telugu though not in Tamil. During their rule from 1676 to 1855 the Maratha kings of Tanjore patronized Ayurvedic Sanskrit and Telugu medical learning, but gave little attention to Tamil medicine which continued to thrive in the villages. The Malayali *vaidyas* employed a compilation of the Sanskrit works of Caraka and Suśruta, while Tamil *vaidyas* on the west coast and elsewhere followed Agastyar and the Siddha tradition. The majority of the southern Telugu medical works were composed in Sanskrit and were either transcriptions of northern Sanskrit texts or ascribed to the ancient sages of south India. According to Ainslie, Tamil medical works were composed in high Tamil or '*Yéllácánum*', that is, *ilakkaṇam*,[9] which suggests that the Tamil tradition was indeed developing separately from the Sanskrit.

Although brahmins practised medicine, with the exception of surgery, according to *Vedic* methods, there was no brahmin monopoly of indigenous medicine in nineteenth-century south India. In *The*

Laws of Manu, medical practitioners were identified as *ambaṣṭhas* and said to be born of a Brahmin man and a Vaishya woman. Because of their Brahmin descent, the *ambaṣṭhas* were held in fairly high regard. The physician caste, the *vaidya*, were shudras[10] and thus prohibited from access to medical knowledge embodied in the brahmanic Vedas. However, in both the Sanskrit and Tamil traditions, the *vaidya* had access to medical text-books (*śāstras*), which were commentaries on the brahmanic texts and believed to be composed by great sages including Caraka, Suśruta and Agastyar. Medical practitioners were found in the Madras presidency among all castes and all religious sects. Some were trained exclusively according to the Sanskritic Ayurvedic texts while others followed the Islamic Unani tradition. Arab medicine contributed to the development of Ayurvedic and Siddha medicine, with practices such as taking the pulse in medical examination, along with the use of mercury in medical treatment, introduced into Ayurvedic medicine from Arabia or Persia. After flourishing in the regions under Mughal suzerainty, Arab medicine also continued in its own right under British rule. In south India, where Muslim invaders made less impact, Arab medicine posed little threat to the Ayurvedic and Siddha systems, though some *hakeems*, practitioners of Unani medicine, did develop a sound medical reputation. A British medical officer reporting on Kurnool in 1843 noted that several *hakeems* resided there 'whose knowledge is obtained from the writings of the old Arabian physicians'. He continued that while they were unaware of the circulation of the blood and 'very ignorant' of anatomy, these physicians 'possessed a copious *materia medica*'. The *hakeems* also occasionally practised bleeding and the removal of cataracts. In addition to followers of both Hindu and Muslim textually based medical traditions, there were folk medical practitioners including midwives, bone setters and supernatural healers who practised their specializations while drawing in part on the humoral theories and magical practices described in the textual medical tradition. Among others, a class of male practitioners restored dislocated limbs, set fractured bones and also practised leech craft.[11]

By the nineteenth century, in Cochin at least, Indian medicine and surgery, which in Siddha medicine included not only operative surgery (*aṟuvai*), but application of heat, including fumigation with vapour (*akkiṉi*), and the application of oil, powder, ointment and paste to remove dead tissue (*karam*), had become distinct hereditary employments. Despite little anatomical knowledge in both the Ayurvedic and Siddha traditions, each offered considerable information on surgical technique and the kinds of instruments employed. Substantial information about surgery in the Siddha tradition is held in the *Ñaṉā nuūl* (religious treatise) attributed to Agastyar.[12] Even so, by the nineteenth century much anatomical knowledge had been lost due to the development of a Hindu prohibition on dissection. Although the British dismissed Hindu scruples as superstition and an impediment to the progress of European science in India, public aversion to post mortems had been similarly powerful in Britain in the early nineteenth-century. The 1832 Anatomy Act, which legalized dissection, was far more an effort to combat the illegal trade in corpses obtained by grave robbing and, during the infamous careers of Burke and Hare, by murder, than a public endorsement of dissection as part of the emerging culture of scientific medicine.[13]

There were also parallels between the practice of surgery in India and England. Operative surgery was not practised by the Telugu or Tamil practitioners in south India principally because it was greatly feared, no doubt due to the pain and risk involved. In England, surgeons had only been recognized as distinct from barbers in 1745, and in the early nineteenth century were still not held in high esteem. As in India, surgery was practised only as a last resort until the introduction of ether in 1846 and chloroform in 1847 made anaesthesia possible and surgery pain-free. In nineteenth-century Indian medical practice, cautery, either by application of caustics or burning of the affected part, was often performed in preference to surgery, particularly in the treatment of tumours, ulcers, abscesses and other skin conditions. The only form of traditional operation performed consistently during the nineteenth century was plastic surgery, particularly Suśruta's technique of nasal reconstruction.[14]

The relationship between indigenous and European medical systems

Similarities

Until the mid-nineteenth-century there was much in common between the European, Hindu and Muslim textual medical traditions. Both European and Indian traditions, which shared early Greek, Roman and, later, Arab influences, understood disease to be a result of the imbalance of humours in the body and the action of 'exciting' and 'predisposing' causes. Reflecting these relationships, there were close structural similarities between Tamil, Greek and Latin medical texts. The Siddha and Ayurvedic methods of diagnosis were similar to traditional European methods of diagnosis by observation of physical signs, based in the teachings of Galen and Hippocrates which, despite developments in diagnostic method, remained the core of nineteenth-century British diagnostics. Both European and Indian traditions, though they rarely cited divine intervention as a direct cause of disease, considered moral conduct to be a significant factor in predisposition to illness.[15]

The similarity between European and Indian medical traditions apparent in the early nineteenth century was reflected in the south Indian conceptualization of leprosy as a disease. According to the *Citta maruttuvam*, based on a treatise of the siddha Yūkimuni, believed to be a follower of the renowned siddha Teraiyar, a close follower of Agastyar, and other Siddha texts, leprosy *peru nōy*, the big disease (Tam), was classified into seven types according to the disorder of the humours, or 18 types according to symptoms, though there was some variation in the naming of each form in the Tamil texts. In the context of the humours, the disorder of 'wind', 'heat' and 'phlegm' resulted in three specific forms of leprosy with four other leprosies caused by the combination of imbalanced humours: 'wind' and 'heat', 'phlegm' and 'heat', 'phlegm' and 'wind'. In both Ayurveda and Siddha, 18 types of leprosy were identified according to symptoms though there was some variation in the particulars of each variety.[16] As with British interpretations of leprosy, there was some confusion as to the correct nomenclature for 'white leprosy' or leucoderma which was sometimes excluded from the list of 18 types.

Other indigenous classifications of leprosy were observed by the British. In South Canara, five forms of leprosy were recognized, and in Chingleput, six. In each case, only one of the conditions classed in indigenous medicine as a form of leprosy actually denoted the specific disease. The other conditions identified as leprosy referred to elephantiasis, psoriasis, leucoderma, syphilis and other skin conditions.[17] Nineteenth-century indigenous classification of leprosy was closer to medieval European concepts of the disease, which had included syphilis and other skin diseases, than to mid-nineteenth-century British notions of leprosy. From the 1840s, and certainly by the 1860s, the British understood leprosy to be a constitutional disease occurring in two distinct types though with a substantial variety of mixed forms.

In the Siddha tradition, the different forms of leprosy were classified principally in terms of the disease's physical manifestation in the skin. The *Citta maruttuvam* for example, identified *jāṉaittōl peru nōy (kaja kuṭṭam)*, 'elephant skin leprosy', *paṉṟitōl peru nōy*, 'pigskin leprosy' and *attikkāy peru nōy (avutumpara kuṭṭam)*, 'fig leprosy' in which 'large and small fig-like tumours shoot up from the skin . . . [and] hang from the body like figs from their stalks'.[18] In addition to alterations and irritation of the skin, other characteristics identified as leprosy by British medical officers in early-nineteenth-century south India were similarly identified by the siddhas. In particular, anaesthesia (often first noticed as insensitivity to fire), ulceration and sores, the decay of fingers and toes, and deformation of digits were recognized by the siddha as signs of leprosy.

As the British tradition linked immorality with leprosy, similarly in both the Ayurvedic and Siddha traditions, leprosy was attributed to bad karma or bad actions in a previous life. Physical causes of the disease were also recognized in both British and indigenous traditions. The *Citta maruttuvam* noted that in addition to 'one's own actions [karma]', leprosy was caused by 'excessive intake of stale fish, crabs, snails and lobsters, consumption of uncooked foods, meditating after taking sumptuous food, close contact with lepers and sleeping on their beds'. Leprosy was also described as hereditary. Yūkimiṉi's *Vaittiya cintāmaṇi* suggested among the physical causes of leprosy 'bilious indolence, coitus with fat women' and eating rice adulterated with husks, bran and dirt. In addition, he identified emotional and moral causes including: 'worries, . . . disrespect to one's teachers, Lord Shiva, Brahmins, deities; desiring another man's wife; betraying trust; swearing at mendicants and indigents [and] rape'.[19]

Like British medical practitioners, the siddhas sought clues to the cause of leprosy in socio-economic, environmental and dietary factors. However, the Siddha placed greater stress on the moral condition of the leprosy sufferer, particularly any form of excess, as contributing to their infection. Poverty, regarded by the British as the principal factor contributing to the incidence of leprosy and linked by some British medical writers to moral turpitude, was not specifically mentioned in Siddha explanations of leprosy causation.

Differences

Although British and Indian medical systems shared a degree of compatibility until the mid-nineteenth century, there were fundamental differences. The most striking difference between the British and Hindu medical systems was the supposedly sacred origin of Ayurvedic and Siddha medicine and their practitioners' customary recourse to divine assistance in effecting cure. This is not to say that a British medical officer never said a private prayer to assist in medical or surgical treatment, but by the nineteenth century, as the last vestiges of medieval culture fell away, the increasing secularization of Europe was evident also in the emergence of a new, often assertively rational, secular scientific medicine. In nineteenth-century south Indian Hindu medical practice, recourse to the gods was an established part of medical endeavour. Before beginning surgery or medical treatment, the gods who superintended the different parts and functions of the body involved were appealed to for aid. Among others, Agni, god of fire who superintended the tongue, Brahmā, the creator, responsible for the soul, and Vishnu, the preserver who watched over life itself, were most frequently invoked. As with the removal of the karmic taint of leprosy required by the *dharmaśāstra*, in the medical treatment of leprosy, reference to the gods was essential. The *Kaṉmakāṇṭam* advised that as the first step in leprosy treatment, 'all karmas have to be expiated', that is, all the deeds in previous lives which contributed to the development of leprosy in the sufferer's body had to be removed.[20]

It was during the nineteenth century that the differences between the indigenous and British medical systems were increasingly perceived by British medical practitioners in India as evidence of the superiority of British medicine. British medical officers in south India found in the religious aspect of indigenous medicine the strongest evidence of its inferiority. Criticism of indigenous medicine as 'unscientific' because of its religious associations was evident even among those who held it in high regard. Whitelaw Ainslie, who professed a deep respect for the professional character of the Indian *vaidya* and committed himself to discovering and publishing a detailed account of Indian *materia medica*, cited the origins of Hindu medicine in sacred Hindu writings to be the principal impediment to a lack of scientific progress in Indian medicine. He considered the religious origins of Hindu medicine to be responsible for the 'state of empirical obscurity in which the science is still sunk in India'.[21] For the medical officer directly confronted with indigenous medicine, its theoretical similarity to European medicine was often less striking than its practical difference. The medical officer for the Coorg found, in 1844, that medicine is in a very rude and simple state; there are no hakeems or persons who practise it exclusively and most if not all diseases are attributed to the influence of an evil eye or the anger of the gods; curative measures principally consist in prayers, incantations and offerings to idols.[22]

Descriptions of Indian medicine as unsystematic, incoherent and unreliable were also common among British medical officers practising in the south in the first part of the nineteenth century. The medical reporter for the Mysore Division of the Madras Army was unimpressed with 'the state of medical science among the natives', noting that:

Of the popular remedies in use many are inert and some are calculated to produce effects altogether different from those for which they are administered; they chiefly consist of aromatic or pungent seeds and gums, with a few mercurial and other mineral preparations, which are extremely rude, and consequently uncertain in their effects.[23]

The view that indigenous medicine was incoherent, essentially mythological and thus, 'unscientific' and inferior to the new scientific medicine of Europe was not limited to medical officers in south India. In 1838 J.R. Martin of the Bengal Medical Service commented that:

The inductive mode of reasoning is unknown to the Brahmins; they have never been observers of common facts: they have no treatises on particular diseases: all they have of record in medicine is in the shape of diffuse general systems, of which the greater part relates more to mythology than medicine.[24]

This view of Indian medicine was characteristic of early-nineteenth century British irritation with Indian culture in general. Similar charges of incoherence and irrationality were levelled at the Hindu legal system by the British during the same period.

British borrowing

Criticism of indigenous medicine in the first part of the nineteenth century, and the notion that British medicine was superior to indigenous, came relatively late in the history of colonial and indigenous medical interaction. There existed in the colonies, not only of British

India but of other expanding trade empires, a long medical tradition of recourse to indigenous medical remedies and to assistance from practitioners of indigenous medicine which dated from as early as the sixteenth century. From the sixteenth to the eighteenth centuries, the Portuguese on the Malabar Coast, the Dutch East India Company and, later, the East India Company, concerned to maintain the health of their employees, were willing to borrow medical remedies from the indigenous medical tradition. The difficulty of obtaining European medicines in India, an expectation that indigenous remedies would be more appropriate for indigenous diseases than European medicines, and the high cost of imported medicines provided strong incentive for the colonial use of indigenous medicines wherever possible. In the early seventeenth century, the East India Company encouraged the use of indigenous remedies among its servants specifically to reduce the need for costly imported drugs.[25]

From their first encounters with India in the sixteenth century, the paucity of European physicians available and the conviction that local physicians were better acquainted with both the diseases and remedies of the region encouraged Europeans in India to resort frequently to local physicians for medical assistance. In the first half of the seventeenth century, the tendency for European surgeons to die as swiftly as their patients resulted in the East India Company making official policy the employment of Indian physicians whenever needed. By the end of the seventeenth century, the assimilation of local medical knowledge and increased European confidence in diagnosis and treatment of local disease had reduced reliance on indigenous practitioners. Even so, European consultation with indigenous practitioners did continue and by the end of the eighteenth century there was renewed interest in Indian medical systems, stimulated by the enquiries of Sir William Jones into ancient Sanskrit medical texts. The East India Company urged its surgeons to learn from the Indian medical texts that were becoming available in translation largely through Jones' efforts.[26]

Official endorsement of Indian medicine in the early nineteenth century was consistent with the wider British policy of non-interference in Indian culture. The British willingness to support and learn from indigenous knowledge was evident particularly in Bengal with the company's establishment of vernacular colleges and medical schools to enable Indian students to study Ayurveda and Unani systems alongside European sciences, including medicine. In the early nineteenth century several books and articles were published by British medical officers in India, reflecting this renewed interest in Indian medicine and the possible benefits it might hold for Europeans there. Ainslie's *Materia Indica* proved popular among British medical practitioners in India, for whom it was principally intended. Despite commenting critically on the lack of empiricism in Indian medicine, Ainslie still considered it to be of value to European medical practitioners.[27]

However, British patronage of indigenous medicine was short-lived. From the mid-1830s the authority of the 'Orientalists', particularly Sir William Jones, declined and notions of utilitarianism and reason gained currency with the rise of the 'Anglicists'. Reflecting also the growing evangelical Christian influence in early-nineteenth-century British culture, the British perception of indigenous medicine as 'superstitious' hastened its fall into disfavour. European medicine, perceived as 'rational' and 'scientific', became regarded as the only legitimate medical system and colonial patronage of indigenous medicine ceased. In 1833, the Grant Committee recommended the abolition of Indian medical colleges and the withdrawal of all support for Indian medicine. The teaching of indigenous medicine in the vernacular was discarded and although medical education in the vernacular was not permanently abolished, only European medicine taught in English received government support.[28]

Despite the increasing representation of indigenous medicine as inferior to the rational and secular medical science emerging in European culture, incorporation of indigenous remedies into British medical practice continued to occur throughout the nineteenth century because of the continuing high cost of imported European medicines.[29] At the close of the nineteenth century, indigenous remedies were still considered essential to British medical practice in south India. The *Civil Medical Code* for the Madras presidency of 1889 stipulated that:

> The indigenous drugs of the Indian Pharmacopoeia are to be used in preference to European substitutes wherever practicable. Imported medicines should not be indented for when indigenous substitutes are readily available. Medical officers should take pains to see that indigenous medicines are collected at the proper season and that they are genuine and perfectly fit for use.[30]

Indian *materia medica*

In the nineteenth century, British investigation of Indian *materia medica* became more systematic than ever before. In the early decades, the gathering of information was fairly *ad hoc*, with little coherent government involvement, pursued principally on the initiative of individual members of the medical service such as H.H. Wilson in Bengal and Whitelaw Ainslie in Madras. While Jones had elucidated the Hindu legal tradition, it was left to these 'scholar-surgeons' who followed to investigate the Indian *materia medica*. In the Madras presidency, from the 1850s, the involvement of government in the co-ordination and administration of the discovery and trial of indigenous remedies as substitutes for expensive British remedies rapidly increased. In order to extend British knowledge of Indian *materia medica* and promote development of the medicinal resources of India, in 1851 the Madras Medical Board instituted committees at various stations of the presidency to obtain, and test for efficacy, indigenous drugs which could be employed as substitutes for those imported from England.[31] The bureaucratization of the relationship between indigenous and British medicine in India was further advanced by the secretary of state in council's commissioning, in 1867, of a pharmacopoeia of India to be edited by Edward John Waring. With the formation of the drug committees, the investigation of indigenous remedies for employment in the British pharmacopoeia was no longer the purview of interested individuals such as Whitelaw Ainslie. The collection and assessment of the efficacy of indigenous remedies was firmly in the hands of government bureaucracy which fostered the institutionalization of indigenous medicine as being of inferior status to European.

The two principal Indian *materia medica* of the nineteenth century, Ainslie's *Materia Indica*, published in 1826 and Waring's *Pharmacopœia of India*, published in 1868, compiled and edited respectively by members of the Madras Medical Service and drawing substantially on the medicinal plants and minerals of southern India, show well the shift in British perspective on indigenous medicine from early to mid-century. Both *materia medica* were intended to make indigenous remedies in India accessible to British medical officers and thereby to reduce British dependence on costly imported drugs. However, the texts differed considerably in their respect for indigenous medicine and confidence in British medical superiority. On the one hand, Whitelaw Ainslie sought 'to supply what has long been wanted, a kind of combining link betwixt the *Materia medica* of Europe and of Asia',[32] building a bridge between two medical cultures, each of intrinsic interest and value. On the other, Waring's work encapsulated the official understanding evident by the second half of the nineteenth century, that the value of Indian medicine lay in its use as a resource for British medicine.

Both texts drew on the existing, substantial but scattered research by British medical officers in India into the therapeutic use of indigenous drugs. Ainslie's work represented the labour of an individual medical officer in the 'Orientalist' tradition, fascinated and humbled by the wealth of indigenous medicine. Waring's *Pharmacopœia* was the result of a systematic, wide-spread investigation sanctioned by the government and prepared under the superintendence of a committee, which included Waring himself and was headed by Sir Ranald Martin, then inspector-general of hospitals, noted above for his considerable lack of regard for indigenous medicine. The *Pharmacopœia of India* was intended as a remodelling of W.B. O'Shaughnessy's 1844 *Bengal Pharmacopœia*, which had been more in the tradition of Ainslie's *Materia Indica*.[33]

In Ainslie's work, the first volume details the usage by the peoples of India and the East of substances present in their region which were also part of the British *materia medica*. The second volume describes medicines 'employed almost exclusively by the Indians and other Oriental nations'. In the first volume, British remedies are presented as part of indigenous medical systems, and in the second the uses of native medicines in indigenous treatment contexts are considered in their own right. By contrast, Waring's *Pharmacopœia* described indigenous medicines principally in the context of British medical practice, assigning them 'official' and 'non-official' status in relation to their use in British medicine. Waring's work proposed a hierarchy of remedies which conferred the highest status of 'official' on all the items already in the *British Pharmacopœia* published by the medical council of England in 1867. The only indigenous remedies discussed were those considered by the Indian Medical Department to be of value as therapeutic agents. 'Non-official' status was given to items considered efficacious in indigenous medicine but either not yet confirmed as such in British medicine or yet to be tested in British medicine but 'deemed worthy of attention'. Indigenous remedies were not incorporated into British medicine on the basis of their reputation in indigenous medicine, but only after tests and trials of efficacy which classified them according to the emerging 'scientific' standards of British medicine. Waring's *Pharmacopœia* represented indigenous medicine as of value only in so far as it contributed to British medicine, and anticipated with satisfaction the formal incorporation of indigenous knowledge into British medicine through educational institutions.[34]

While the teaching of European medicine in juxtaposition with Indian, under British patronage before 1835, was intended to show by contrast the superiority of the European system to the medical students,[35] it also indicated that the government recognized that indigenous medicine had an existence separate from British. By the 1860s, however, the absorption into British medicine of selected indigenous items, tested for efficacy according to the emerging scientific principles of British medicine, suggested that indigenous medical systems had no intrinsic value in British eyes. This shift reflected not only a hardening of attitude towards indigenous medicine from the 1860s, but also a tendency to look towards Europe for medical guidance rather than to indigenous medical culture, as had occurred at least until the end of the eighteenth century. The turn towards Europe also brought the medical investigation of leprosy increasingly under the influence of the developing European scientific culture and of British medical and governmental authorities.

4
Leprosy Treatment: Indigenous and British Approaches

In the British medical tradition leprosy was believed to be incurable and, despite the promise of gurjon oil and other late-nineteenth-century treatments, it remained so until the discovery of Dapsone in the 1940s. The lack of any specific British treatment for leprosy was, particularly prior to the publication of Danielssen and Boeck's findings in 1848, partly due to uncertainty as to what actually constituted the disease, and, until Hansen's discovery of the leprosy bacillus in 1875, partly to ignorance of its true cause. In the Siddha tradition, however, some forms of leprosy were regarded as curable, probably because some of the diseases classified as leprosy were not forms of leprosy at all and were responsive to Siddha treatments. The *Kanmakāntam*, attributed to Agastyar, advised the leprosy sufferer that if he or she 'listens to the doctor and does all that he says, death can be averted'. According to the *Citta maruttuvam*, ten forms of leprosy were incurable, while 'the rest may be cured with proper treatment, medication and the disciplining of body and mind'.[1]

Neither the British nor the Indian traditions had any specific medicinal cure for leprosy. Rather, both employed treatments used for a range of conditions regarded as skin diseases, especially syphilis. The clear separation of leprosy from venereal disease did not occur in either Europe or India until well into the nineteenth century. It was not until the early twentieth century that the presence of acid-fast *Mycobacterium leprae* was identified, finally allowing a bacterial diagnosis of leprosy. This was a valuable adjunct to diagnosis of cutaneous symptoms and could prevent misdiagnosis of venereal and other diseases as leprosy. The hope of discovering a specific cure for leprosy did exist among British medical practitioners in early-nineteenth-century Madras. Dalton, in 1813, expressed his hope that even if his proposed cure ultimately proved ineffective, 'a preferable remedy to the one that I had the happiness of selecting will yet be produced'.[2] However, the search for a specific cure did not become a significant concern for either the British medical profession (both in India and England) or the Indian and British governments until the 1870s.

There was much similarity between nineteenth-century British and Indian approaches to the medical treatment of leprosy. In both traditions an initial purgative was often given to cleanse the body and prepare it for treatment, followed by various therapeutic regimens such as dosage with oil and tablet formulations, the external application of ointments, the fumigation of the leprosy sufferer with various types of smoke, and prescription of a specific diet.[3] Both the British and Indian medical systems emphasized the importance of physical cleansing in the treatment of leprosy, although among leprosy sufferers in south India washing was often popularly believed to be a stimulus to the disease. Assistant Surgeon Paul noted in 1855 that at the Madras Leper Hospital:

Miserable objects in every degree of loathsome wretchedness are admitted covered with, or rather encrusted in, filth – indeed many had not washed for years in the belief that ablution aggravates the disease, but after the plentiful use of soap and cold water daily for a time, their sores heal, their skin gets more healthy, and they even gain flesh.[4]

Such beliefs among the leprosy sufferers who presented themselves at the Madras Leper Hospital for treatment might have been due to ignorance of indigenous medical ideas of cleanliness. Indian medical treatment was often expensive, so that the poor frequently resorted to European institutions where treatment was free, even though they might have preferred indigenous remedies.[5]

In the Hindu tradition, daily cleansing of the body was as much a matter of religious observance as an aspect of medical practice. In south India, bathing had a specific role in many, if not all, forms of prescribed treatments for leprosy. The *Carapēntira vaitiya muṟaikaḷ*, a Tamil medical text composed between 1802 and 1832, dealing with gout, rheumatism, piles, leprosy and disorders of bile, mentions applying a paste of ground *vāluḻuvai* ('staff tree' oil) rice and cow's milk to the body and then removing it by bathing as part of a cure consisting otherwise of consumption of the same mixture and of raw peanuts. In a more complex, 40-day course of treatment, taking an oil bath every ten days was stipulated. The Siddha text, *Akattiyar paḷḷu*, also included bathing as a specific aspect of fumigation treatment. In one treatment, after fumigation of the leprosy sufferer with smoke seven times in three-and-a-half days, and the repetition of the same procedure under diet restrictions for another three-and-a-half, the patient was required to take a hot-water bath on the eighth day.[6]

Medical officers employed in the care of leprosy sufferers at the Native Infirmary and later the Leper Hospital, were generally in agreement as to the benefits of physical cleanliness in markedly improving the condition of leprosy sufferers, though they recognized that it could not prevent the progress of the disease. Even so, what precisely constituted cleanliness in British terms was still open to debate. The germ theory of disease causation which informs the modern medical concern with hygiene and sterile surgical conditions, was not established until the 1870s and 1880s. Dalton's recommendation in 1811 that a combination of Madeira wine and the urine of a healthy person be used to cleanse leprous ulcers was condemned by van-Someren as being based on an erroneous understanding of 'cleanliness'.[7]

Despite the differences between the European and indigenous traditions, throughout the nineteenth century British leprosy treatment was very much a matter of employing the most effective remedies available, no matter which tradition they came from. From the 1860s, British concepts of leprosy and leprosy treatment were increasingly privileged over indigenous ones by the government and medical authorities in Britain and by the Indian government. The British government and medical profession had come to assume the superiority of a British 'scientific' medical approach to leprosy. However, at the local level the limited effectiveness of leprosy treatments sanctioned by the British and Indian governments was evident, and indigenous treatments continued to be employed by practitioners of British medicine in India.

Remedies

The principal treatments attempted up to 1861 in the Madras Leper Hospital included both British remedies, such as Donovan's Solution and Fowler's Solution, and exclusively indigenous remedies such as the Asiatic Pill. The notable exception was the bark and oil of the *Melia azadirachta*, in Tamil *vempu*, commonly known as *neem*, from Hindi, or *margosa*, from the Portuguese word for 'bitter'. In Portuguese Goa, daily massage of the body with sap from the neem tree was a common indigenous treatment for leprosy. Where sap was not available, the patient took a paste daily for 40 days which was formed of 5 gms of pepper ground together with 10 gms of neem leaves.[8] Although familiar to the British as a tonic in the indigenous *materia medica*, *neem* was not well known as an aspect of indigenous leprosy treatment until the end of the nineteenth century.[9] Agastyar claimed that when one medicated oil treatment for leprosy based on *neem* was applied to the sores, 'the scales will fall off; the skin will glow like gold'.[10] In addition to the frequent use of *neem* bark as an internal remedy for leprosy, in at least one Siddha fumigation treatment, *neem* oil was used to prepare the wick before lighting.[11]

In Unani medicine neem was also an important leprosy treatment. In his 'translation' and considerable emendation of the *Taleef Shereef*, Playfair cites its use by Unani physicians in treatment of *juzam*, leprosy, and the intractable, foul ulcers which often accompany the disease. In the *thathu virthi bodhini*, a Tamil Ūnani treatise dated 1893 which was primarily concerned with enhancing sexual pleasure and sexual health, a treatment composed of seven parts of the *neem* plant was recommended for the cure of leprosy. The author noted that the prescription was learnt from an old man and had proved to be effective, completely extirpating the disease if taken in the initial stages.[12]

Most remedies used at the Madras Leper Hospital were formulations of arsenic and mercury, substances with a long history of usage in both Indian and European medicine in the treatment of leprosy, venereal and skin diseases, reflecting the close association of leprosy and venereal disease in both traditions. The best insight into the leprosy sufferer's experience of treatment with arsenic and mercury preparations is offered by the alleged cure developed by Dalton in the first decade of the nineteenth century at the Madras Native Infirmary. Dalton's treatment of leprosy illustrates well the tendency to treat venereal disease and leprosy as if they were part of the same disease complex. Specific aspects of Dalton's treatment, particularly the deliberate stimulation of intense perspiration and use of mercury, were typical of sixteenth-century European treatments of syphilis which had changed little by the nineteenth century.

There is no existing record of the actual compound employed by Dalton. However, the instructions for treatment suggest that its active ingredient was a mercurial preparation, probably bichloride of mercury. Though attempted in good faith, the treatment was not only ineffective but also offensive, unpleasant and complicated for the patient. The leprosy sufferer was to take one pill containing the relevant compound morning and night 'until the mouth becomes sore and the gums spongey', whereupon the pills were to be discontinued and the mouth frequently washed and the throat gargled with a mixture of alumen, honey, opiate and decoction of Peruvian bark or *cinchona* (Peruv). When the soreness of the mouth had passed, the pills were to be resumed and the process of alternating remedies continued until every symptom of leprosy disappeared. In addition to making eating painful and difficult, the regime required the limitation of the leprosy sufferer's diet to 'food of easy digestion', excluding vegetables, pork and fish and allowing only minimal spice. To the discomfort of the treatment was thus added a diet not only less nutritious but also with a considerable reduction in spice, the essential ingredient of flavour and one of the few items to relieve the monotony of the leprosy sufferer's daily institutional life. Further, Dalton suggested that it might be necessary to employ castor oil or some other, equally mild purgative during the course of treatment.

The external symptoms associated with the disease were also attended to. Excoriations on the body and legs were to be rubbed with camphor ointment twice a day and ulcers were to be cleansed with the lotion recommended by Dalton based on Madeira wine and urine. After cleansing, the ulcers were to be dressed twice a day with balsam of copaiba and, where this was found to be ineffective, with a mild solution of blue stone (copper sulphate) in water. To the offensiveness of being washed with human urine, particularly objectionable to Hindus as a ritually impure substance, was added further physical discomfort. Those undergoing treatment were required to be clothed to the extent that the leprosy sufferer 'may always, if possible, be in a perspirable state, as it is very desirable that the medicine should operate in this way'. Further, the usual ablutions which made the hot and humid climate tolerable in Madras, such as washing the head and rubbing the body and head with oil, were forbidden for fear of instigating fever or 'severe flux'.[13]

Dalton's treatment, though employing mercury which was used in Indian treatments of syphilis but not regarded as a specific for leprosy, was essentially European, employing substances of the British *materia medica*. The treatment was designed to address both the constitutional and external manifestations of leprosy and to ameliorate the effects of the mercury treatment itself. Bichloride of mercury is a corrosive poison, far more active than pure mercury, which was understood in the nineteenth century to 'impoverish' the blood by reducing its content of red corpuscles. A typical consequence of both internal and external treatment with mercury was the condition of mercurial ptyalism, characterized by swelling and ulceration of the gums, odorous breath and excessive salivation, which signified that the body had become saturated with mercury. The mouthwash and gargle adopted alternately with the mercurial preparation was formulated to alleviate the symptoms of mercurial ptyalism. Alumen acted as an astringent, reducing swelling in the gums and the opiate reduced pain. Peruvian bark, which by 1800 was commonly taken internally as an antiseptic in the treatment of gangrene and mortification, used here as a mouth-wash, alleviated the ulceration and sponginess of the gums. Honey was added as a combinant and to make the entire mixture palatable. Camphor ointment was an external application typically used to ease the joint pain of rheumatism, and to both toughen and soothe irritated and itching skin. Van-Someren's objection that the combination of Madeira wine and the urine of a healthy person as a cleansing lotion was manifestly unhygienic suggests that urine was, by the mid-nineteenth century at least, not a typical cleanser in European medicine. In the nineteenth century, wine, like Peruvian bark, was typically employed as an internal antiseptic to treat tissue degradation. Wine had a long history in medicine, both taken internally as a cordial and a tonic, and as an external cleanser for wounds. Balsam of copaiba, imported from Brazil, was a common British treatment for gonorrhoea though usually taken internally. Blue stone (copper sulphate) was similarly employed in the treatment of gonorrhoea and the ulceration and sores associated with skin disease.[14]

There is no direct information on the response of leprosy patients to Dalton's efforts to effect their cure. It is clear, however, that among leprosy sufferers there was a strong desire for relief from the disease. The rumour of the effectiveness of Dalton's cure in 1813 was sufficient to cause a sudden increase in admissions of leprosy sufferers to the Monegar Choultry and Native Infirmary in Madras. It is not clear whether or not force was employed to ensure the leprosy sufferers' co-operation in treatment though the fact that leprosy sufferers were able to leave the asylum while undergoing Dalton's regimen suggests that compulsion was not used. Once in the hospital, it is more likely that the hope of cure gave those with leprosy the courage to persevere with the revolting and highly dangerous treatment. Many patients subjected to repeated courses of mercury poisoning died quickly without cure of their disease. However, Dalton's boast of the power of his remedy 'to heal its dreadful ulcers – stop the progress of joints falling off, restore sensation where there was no feeling and relieve the unhappy sufferer so far as to render his existence comfortable to himself and innocuous to others'[15] suggests that at least the ulceration and damage to the skin caused by leprosy were soothed and healed, presumably as a result of the external applications of oils.

Dalton's treatment, though taxing, painful and unpleasant for the leprosy sufferer, was no worse than British medical treatments such as purging, bleeding and mercurial remedies which were commonly employed in the treatment of British soldiers in India. Leprosy sufferers, who were predominantly Indian and Eurasian, were not subjected to medical treatment any more harsh than that given to the soldiers who were often their social and economic superiors. The British soldier endured mercurial treatments for his venereal disease, as did the non-European leprosy sufferer for leprosy. It was a sign of the desperation of British medical officers to find effective treatments for leprosy sufferers that mercury was employed in treating Indian patients at all. In the early nineteenth century, bleeding and mercury treatments were generally not considered appropriate for the Indian population in the belief that they were less able than the British to withstand physical hardship. Rather than an example of racism expressed through colonial medical practice, British use of mercury in leprosy treatment ran contrary to the prevailing perception of the Indian as physiologically weak. The arguments against mercury treatment in leprosy raised in the 1860s and again in the context of Beuperthuy's cure, were based on the effect of mercury on the constitution of an individual suffering leprosy, not on ideas of racial weakness. Further, despite the medieval European associations of leprosy with sexual immorality and the continued link between venereal disease and leprosy into the nineteenth century, the Madras records do not suggest that mercurial treatments for leprosy were perceived by the British as a form of punishment for sexual excess.

News of both indigenous and European leprosy treatments employed by the British in India, such as chaulmugra oil, biniodide of mercury ointment and arsenious acid, reached the British medical press. Some common European remedies such as potassium iodide and hydriodate of potassium, still used in the treatment of rare cases of leprosy occurring in nineteenth-century England, were given shorter shrift in India. F. Taylor of Woodstock reported to the *Lancet* in 1866 the great benefit to a sufferer of 'lepra syphilitica' from treatment with iodide of potassium and *liquor potassae* taken internally, together with the external application of diluted nitrate of mercury ointment. In Madras, Hydriodate of Potass was noted by Surgeon Evans in his 1847 report for the Madras Leper Hospital as being at most of 'at least temporary benefit' in only one or two cases of leprosy.[16]

Fortunately for the leprosy sufferers of Madras, not all British remedies for leprosy were employed in India. In 1860, the *Lancet* noted the recommendation by several medical writers of employing bleeding in addition to the usual mercurial regimens in the treatment of leprosy and other skin diseases. The author noted that although bleeding was rarely practised, he had found rapid improvement in two women believed to be suffering leprosy who were treated by bleeding in addition to dosage with various forms of potassium.[17] He described one as possessing the 'syphlitic' form, reflecting the continuing confusion of leprosy and syphilis in Britain.

Tonics and palliatives

As van-Someren noted, some of the remedies attempted at the Madras Leper Hospital were inert and did the patient no actual harm, while others such as Dalton's cure were dangerous, 'acting as irritant poisons, and only seemed to reduce the strength and powers of resistance to disease' in the leprosy sufferer. The Asiatic Pill, based on arsenic, a strong poison, brought minor improvement in the external symptoms of leprosy but also caused severe intestinal and cutaneous irritation. Although there was some improvement to the patient's well-being and the condition of their skin once the irritation subsided, the longer-term effect of the substance was to cause 'a general fading of the system'. Similarly, Donovan's solution and chaulmugra oil were found to be of no significant benefit to the leprosy sufferer.

Since medical intervention, in particular the use of mercury, had been found to be detrimental to the leprosy sufferer's health rather than offering effective cure, van-Someren recommended recourse to the *materia medica* for substances which acted as tonics and palliatives in preference to strenuous cures. He suggested cod liver oil, chalybeates, (water impregnated with iron) and preparations of iodine and iron as internal remedies to improve the quality of the blood. External measures included the use of sulphur vapour-baths, the application of emollient oils to ease dry and split skin, and of calamine cerate, astringent lotions, water dressing or cataplasms (poultices) to sores according to the needs of each case. Because of the leprosy patient's poor general health, van-Someren also recommended against strenuous medical treatment of the other diseases such as dysentery, diarrhoea, albuminuria and pulmonary affectations to which leprosy sufferers were highly susceptible.

Other British medical officers at the hospital also recognized that, despite the ineffectiveness of the available remedies for leprosy in both the British and indigenous medical traditions, much could be done to mitigate the most severe symptoms of the disease and to make life more comfortable for the leprosy patient. Van-Someren endorsed the testimony of the other medical officers, asserting that considerable improvement in the conditions of the leprosy sufferers could be secured by the benefits of hygienic conditions: 'Good food, pure air, a rigid attention to cleanliness, and a certain amount of bodily exercise, certainly contribute more than anything else to ameliorate the health of lepers'.[18]

Fumigation

One treatment for leprosy employed by the British which enabled intensive cleansing of the skin through perspiration, yet spared the leprosy sufferer the severe consequences of mercurial treatment, was fumigation. In indigenous medicine, fumigation was a highly developed technique of treatment for leprosy, genital cancers and ulceration. The *Akattiyar paḷḷu* suggested several smoking techniques. One recommended a smoking preparation combining well-cleaned zinc, subacetate of copper, white arsenic, red and yellow orpiment, powdered cobra's meat, red sulphate of mercury, sesame and honey-wax in measures of two and a half gold coins. To these substances were added cotton-leaf extract and the whole mixture was ground together. The directions for fumigation were as follows:

> Make the patient sit on a chair and drape him up to the neck in thick canvas/cloth. Burn tamarind twigs and put one piece of the above preparation in powdered form under the chair. If this

is done, the smoke of the medicine will easily spread over the whole body. Take care not to expose the eyes and face.[19]

In other smoking treatments for leprosy, the smoke was inhaled through a bamboo pipe or directed specifically onto the sores.[20]

Fumigation with carbolic acid vapour was practised at the Palliport and Calicut asylums by Deputy Surgeon-General W. Johnston. Fumigation, often with mercury or arsenic, was a common European treatment for venereal disease until the sixteenth century and was still employed, though less frequently, by the nineteenth century. Johnston's trial of carbolic acid vapour fumigation, however, appears to be the only application of the technique by the British in south India in the nineteenth century, although in 1855 the French had tested fumigation with *Hydrocotyle asiatica* as a leprosy treatment in Pondicherry.[21]

Johnston's use of carbolic vapour was consistent with its application in surgery and the more general treatment of disease developing in Europe. From the mid-1860s, Lister employed carbolic acid in his pioneering work in antiseptic surgery, and in India a strong endorsement of the effectiveness of carbolic in killing 'all septic germs' thus removing 'many *causes* of disease [sic]' was reprinted in the *Indian Medical Gazette* from the British publication, the *Pharmaceutical Journal*. Carbolic acid was part of the British pharmacopoeia by the mid-nineteenth century and was thus listed as an 'official' remedy in Waring's *Pharmacopœia of India*, which described it as a powerful deodorizing and disinfecting agent, valuable in the treatment of obstinate ulceration and chronic skin disease.[22] In the early 1870s carbolic acid was occasionally employed by the British in India as an internal remedy for leprosy, but without success. More commonly, carbolic acid ointment was rubbed into all parts of the body and in some cases applied neat to leprous patches, significantly reducing the external symptoms of the disease. The remedy was not unanimously held in high regard as a leprosy treatment in England. The *Lancet* observed that carbolic acid would prove of little benefit in leprosy without the support of good hygiene and diet. Professor Hebra of the General Hospital, Vienna, was less restrained in his criticism. He was reported in the *Lancet* of 18 March 1871 as commenting that the only use of carbolic oil in leprosy treatment is to make the patients 'stink' and that, if rubbed with cod-liver oil or any fat, they would be just as benefited![23]

At the Palliport Lazaretto, leprosy sufferers were individually fumigated in baskets until, in August 1875, 'a cage was made of mats and bamboos, covered with painted cloth, by which 10 are fumigated together'. In the cage, the patients sit 'back to back, and are fumigated by two stills, one at each end of the cage, from which carbolic acid with water is vaporized, and they breathe by bamboos protruding from the basket'. W. Doyle, civil surgeon, commented to the deputy surgeon-general that, although the hospital assistant had reported that leprosy sufferers 'have to be almost dragged to the fumigating basket', fumigation was not as unpopular as the other forms of treatment involving dosage and external application of cashew nut and gurjon oils. Doyle suggested, in his annual report, that the treatment was preferred by many because 'it is a vapour bath which they can enjoy gregariously, 10 at a time, and at their ease'.[24] At the Calicut Leper Hospital, fumigation was performed using a cone in which the patients could sit, and a bell tent was also available for fumigating specific sections of the body. A private of the Native Infantry whose left arm showed signs of leprosy was treated using the bell tent, his arm being inserted in the tent and exposed to intense carbolic acid vapour three times a day.

Fumigation with vaporized carbolic acid had been found by the British to have a beneficial effect on the skin, softening and breaking down tubercules and restoring its normal colour and appearance. It was believed that fumigation restored feeling to areas of anaesthesia, though loss of feeling in specific nerves infected with the leprosy bacillus is now known to be irreversible. Further, Johnston, the initiator of the fumigation experiments, considered that the effect of daily copious perspiration contributed to the leprosy sufferer's health by removing all foreign impurity from the surface of the skin, stimulating and restoring the flow of nutrition to the skin and reviving the skin's role in excretion.[25]

Surgery

From the late 1870s, in addition to medicinal approaches to leprosy, the application of European surgical technique to alleviate leprous symptoms became increasingly common in north and western India, though not in the Madras presidency. It was not only leprosy patients in the Madras presidency who did not receive surgical treatment available in other parts of India. Although surgical operations were performed in the hospitals and dispensaries of the Madras presidency, Madras lagged far behind Bombay, the North-Western Provinces and the Punjab in the provision of surgical relief to Indian patients. W.R. Cornish, then surgeon-general with the

Madras government, reporting on the civil hospitals and dispensaries for the year 1880, attributed this neglect to the difference in medical staffing between Bombay, the North-Western Provinces, the Punjab and the Madras presidency. In the northern and western administrations, many of those staffing the civil hospitals and dispensaries were highly trained and qualified assistant surgeons, whereas in Madras a large proportion of the dispensaries were staffed by men of the lowest grade of the Subordinate Medical Service who lacked any education in surgery. In Madras, even the civil surgeons performed little surgery because of 'lack of anatomical expertise'. Cornish observed that those educated in Europe or Asia were equally poorly trained in either anatomy or surgery.[26] This neglect was evident in the dearth of surgical treatment of leprosy in the Madras presidency. In the Punjab and Bengal, British surgical methods of alleviating the symptoms of leprosy such as amputation of toes, feet and legs were increasingly practised from the late 1870s. In one case, the deeply ulcerated foot of a pilgrim from Juggernath was removed by the civil surgeon at Hazaribagh and a prosthetic foot specially constructed in its place. In 1883, the case of a leprosy sufferer whose leg had been removed by amputation was presented to the Calcutta Medical Society in order to demonstrate that even in someone with leprosy who was thus susceptible to ulceration, surgical wounds could heal well, a finding endorsed by the other medical officers present.

By far the most common surgery performed on leprosy sufferers was stretching of the ulnar and sciatic nerves.[27] The procedure was first reported in the *Indian Medical Gazette* of July 1877 as a treatment to alleviate neuralgic pain and restore the loss of sensation in arms and legs occurring in cases of anaesthetic leprosy.[28] The application to the symptoms of anaesthetic leprosy of J.N. von Nussbaum's technique of nerve stretching to relieve neuralgia was proposed by Professor McLeod of Calcutta. The procedure was intended to act 'as a means of freeing the axes cylinders of the nerves from the compression and constriction they undergo by a thickening of their connective tissue elements, or by a deposit within their sheaths',[29] aspects of leprosy recognized and treated today with surgery and drug therapy. The simple operation was performed under antiseptic conditions, usually under 'carbolic spray', on patients anaesthetized with chloroform. The arm was incised to expose the ulnar nerve which was then lifted, either on the back of the surgeon's knife or hooked over a finger, and stretched intensively. The same technique was applied to the sciatic nerve in the leg. Though generally highly regarded as a safe and effective means of easing anaesthetic symptoms, G.C. Roy of the Bengal Medical Service offered a caution, arguing that 'local operative interference' as a remedy for a constitutional disorder was unsound. Roy quite correctly observed that while nerve-stretching may alleviate symptoms it did not remove the constitutional basis of the disease.[30]

In terms of the relationship between indigenous and British medicine, the performance of surgery upon leprosy sufferers was an area in which British medicine could claim superiority over indigenous with some credibility. Even though British internal remedies for leprosy were no more effective than indigenous ones, in the area of surgery there was the opportunity for the Madras surgeons to follow the practice of the northern medical services and provide a more dramatic alleviation of leprous symptoms and disabilities by surgical intervention. By the nineteenth century, surgery was rarely, if ever, practised within the indigenous medical systems, largely because it was dreaded. As a consequence, the provision of surgery was believed by the British to be the most effective means of convincing the Indian populace of the superiority of British medicine and, thus, of the advantages of British rule. Cornish advised the Madras government that the neglect of surgery in the civil hospitals and dispensaries of Madras was a lost opportunity to display British medical superiority. He argued that 'there can be no doubt that the people of India estimate European medical treatment more by what they see of the skill of our surgeons than by any professed faith in the value of our treatment of internal disease'. Even so, despite the opportunity to shore up Indian belief in 'the advantages and superiority of European methods' of surgery,[31] it is unlikely that surgical practice among leprosy sufferers made much impact on indigenous perceptions of British medicine. There were too few operations performed on leprosy sufferers, especially in Madras, and those who did receive surgical attention were most likely among the poorest, least influential members of Indian society.

Gurjon oil

The discovery of a specific cure for leprosy did not appear likely until the early 1870s when, in 1872, Beauperthuy's treatment, involving dosage with mercury and the external application of cashew nut oil, came to light in Venezuela.[32] The gurjon oil treatment for leprosy proposed by Surgeon Dougall, an officer of the Madras Medical

Service, was the first instance of a leprosy cure developed by the British from materials available in the Indian medical tradition. Dougall was medical officer from March 1873 at Port Blair, a penal settlement which included a 'barracks' for leprosy sufferers, situated on the Andaman Islands and under Government of India jurisdiction. Dougall had found the requirements of Beauperthuy's treatment, particularly in respect to diet and hygiene, impossible to fulfil in Port Blair. Motivated by concern for his patients, Dougall preferred to try a more practical means of cure, employing inexpensive remedies obtainable in the settlement.[33]

Gurjon oil, commonly known as gurjon balsam or wood oil, was derived from the *Dipterocarpus turbinatus* tree. Dougall's *Dipterocarpus lævis* was an abbreviated form of *Dipterocarpus indicus, lævis*, the alias for *Dipterocarpus turbinatus*. The British were aware of the use of gurjon oil in indigenous medicine both internally and externally as a stimulant and diuretic and as a common treatment of gonorrhoea. On the evidence available it seems possible that gurjon oil was used in Ayurvedic treatment of leprosy,[34] though it certainly did not have the same reputation as chaulmugra and neem. Since its value as a substitute for copaiba in the treatment of gonorrhoea was noted in 1838 by W. O'Shaughnessy, who later compiled the *Bengal Pharmacopœia* and was particularly interested in investigating indigenous remedies at the Bengal dispensary, gurjon or wood oil had been used by the British in India as a cheap, local substitute for their usual remedy, cobaiba, in the treatment of gonorrhoea. The only use for gurjon oil in either indigenous or British medicine with which Dougall was familiar was this use as a substitute for copaiba.[35] The most common uses of gurjon oil were not medicinal but commercial. Gurjon was used as a natural varnish, a substitute for tar in sealing wooden seams in ships and for preserving timber from white ants. Even so, the gurjon oil sold in the Madras bazaars as paint or varnish could be employed medicinally in the same way as the oil obtained from the government store.[36]

Like the earlier efforts at the Madras Leper Hospital to discover a cure for leprosy, Dougall's treatment was principally a local, personally motivated response to the pitiful medical condition of those under his care. In his *Report on the Treatment of Leprosy with Gurjon Oil*, Dougall described some of the cases of leprosy which confronted him on his first visit to the barracks where the leprosy sufferers were housed:

> In one corner of the ward a man named Gunga Ram was lying
> covered up by his blanket, and when he moved a swarm of flies
> rose with a 'buzz;' this man had been in this condition for two
> years, just able to crawl from his corner at meal times and for
> other necessary purposes, which being attended to, he returned
> and resumed his wonted place and position in order to protect
> himself from his constant tormentors, the flies . . . the whole of
> the men had that blank hopeless look which conveyed to my
> mind the idea of a living death.[37]

At the time of Dougall's arrival in March 1873, the diet allowance
to the leprosy sufferers held in the penal colony was minimal, of
little benefit to people suffering from a disease in which, as Dougall
commented, 'good living is deemed an essential element in every
mode of treatment yet adopted or advised'. The medical treatment
received consisted of the application to leprous sores of carbolized
oil, regarded by some medical authorities in Europe as worse than
useless in the treatment of leprosy, and a daily bath, in which the
leprosy sufferers were washed in water with bran used as a deter-
gent. As Dougall observed, the treatment did nothing to relieve
the patients from the persecution of flies attracted to their sores.[38]
  Dougall described in some detail the leprosy sufferer's day under
the new gurjon oil treatment. The patients, all male, rose at dawn,
gathered some dry, finely pulverized earth and used it to wash in a
nearby stream. Dougall had stipulated the use of earth as more
effective in removing the residue of gurjon oil than either bran or
soap. Possibly intentionally, this change fitted well with Indian ideas
of cleanliness, since earth was used in the Hindu tradition for both
ritual and physical cleansing.[39] The patients returned to the ward
by 7:00 a.m., where, sitting in a row, each man was given his morning
four-drachm dose ($\frac{1}{2}$ oz /14g)of gurjon oil and lime water mixed in
equal portions, which he swallowed before the apothecary. Each
man was then given gurjon ointment in a small vessel, usually
half a coconut shell. The patients then proceeded to rub themselves all over with the oil, keeping up a continuous and thorough
friction for two hours. Meanwhile, the compounder (an army or
government pharmacist) walked up and down the line, 'armed with
a tin of ointment and a spatula' for distributing more ointment as
it was required amongst the patients. No limit was placed upon
the amount given, other than that whatever was given 'must be
well rubbed in'. After this first session of rubbing, the men per-

formed various tasks around the grounds then were required to 'fall in' and swallow their second four-drachm dose of gurjon oil. Ointment was again supplied for external application, the rubbing being kept up from 3:00 to 5:00 p.m.

Dougall was enthusiastic about the value of his new treatment. Unlike Beauperthuy's cashew nut oil regimen, the external application and massage of gurjon oil caused neither pain and blistering nor smelled offensive. The treatment of external vigorous and prolonged rubbing was insisted upon not only for the beneficial action of the ointment on the skin but, in combination with some other gentle employment in caring for the grounds, also for its mental benefits. Dougall was convinced of the value of such exercise to the leprosy sufferer's mind, since through it, their attention was 'being withdrawn from their former hopeless condition and they are let to compare the present with their past'. Dougall took obvious delight in seeing 'the will with which men, who for years had not been able to handle a walking stick on account of loss of fingers and numbness in the arms, now set about this sort of work', and their pride in being able to work again 'when such a hope had been abandoned long ago'.[40]

However, from the mid-1870s until the introduction of sulphone drugs in the 1940s, it was not gurjon oil but the indigenous remedy chaulmugra which became an increasingly important element in British and other European leprosy treatments. In Ayurvedic medicine, chaulmugra seeds, freed of integument, were beaten together with ghee to form a soft mass and applied externally in cases of scrofula, rheumatism, leprosy and other skin diseases. The oil pressed from the seeds was taken internally as well as being applied topically and used as a cleanser. Dosage with the kernel of the chaulmugra seed was considered a potent remedy in both leprosy and diabetes. Exclusion of salt-meats, acids, spices and sweet-meats, and the addition of oily foods, ghee and butter typically accompanied indigenous treatment with chaulmugra.[41]

Throughout the nineteenth century, British and European investigators of chaulmugra mistakenly believed that the indigenous chaulmugra oil consisted of the expressed oil of the seeds of the tree *Gynocardia odorata*. The origin of the British error was to be found in Dr William Roxburgh's *Hortus Bengalensis*, a catalogue of the plants growing in the East India Company's Botanic Gardens at Calcutta, published in 1815. Roxburgh, a surgeon and naturalist, thought that the seeds of the 'kalaw' tree, believed in Burmese

Buddhist legend to have cured of leprosy both Rama, once king of Benares, and the princess Priya, were those of the tree *Chaulmoogra odorata* under another name. In 1819 the same plant was described under the name *Gynocardia odorata* in Roxburgh's *Plants of the Coast of Coramandel,* and the error continued to be reproduced until 1900 when Colonel David Prain, director of the Botanical Survey of India, identified 'the real chaulmoogra seeds of the Calcutta bazaar and of the Paris and London drug dealers' as *Taraktogenos kurzii*.[42]

In his note on chaulmugra, published in 1854, F.J. Mouat of the Bengal Medical Service endorsed Roxburgh's mistaken identification of the source of chaulmugra as *Chaulmugra odorata* in another of his works, the *Flora Indica,* and the same plant's subsequent description under the name *Gynocardia odorata* by other investigators. Waring listed *Gynocardia odorata* as 'official' in the *Pharmacopœia of India* of 1867 and named it as the only source of chaulmugra oil. Maclean's *Glossary,* published some years later in 1893, mentioned *Hydnocarpus odoratus* as an alias of *Gynocardia odorata* and, similarly noted *Gynocardia odorata* as the only source of chaulmugra oil.[43]

Chaulmugra's mistaken identity reflected the gap between official British understanding of indigenous remedies and the knowledge and use of indigenous remedies employed in leprosy treatment at the local level. No matter how chaulmugra was officially identified, the actual substance available as an indigenous remedy for leprosy was true chaulmugra or a close relation. The chaulmugra mentioned in the Ayurvedic text *Suśruta Saṃhitā* and available in south India was obtained not from *Gynocardia odorata* but from the *Tuvaraka* (Skt) or *Hydnocarpus wightiana* plant of the south-west coast, known as 'chaulmugra' in Hindi and Persian. It was a close relative of the *Taraktogenous kurzii* plant of Assam and Burma which provided the chaulmugra oil for the Calcutta bazaars. Maclean's *Glossary,* on the basis of research in the 1870s, stated that the active principle in the treatment of leprosy was gynocardic acid, found in the *Gynocardia odorata.* However, as researches at the beginning of the twentieth century have shown, chaulmugric acid, not gynocardic acid, was the active principle in leprosy treatment and the former was not contained in *Gynocardia odorata* at all. *Gynocardia odorata* was noted as one of the sources of chaulmugra oil in the *British Pharmacopœia* of 1898 and it was not until the 1914 edition that this publication recognized that oil containing chaulmugra acid, the active principle in the treatment of leprosy found in *Hydnocarpus wightiana,* could be obtained from *Taraktogenous kurzii*.[44]

Better identification of the true indigenous treatments for leprosy among the British medical officers in Madras no doubt contributed to their greater respect for indigenous remedies. In south India, oil from *Hydnocarpus venenata* or *Hydnocarpus inebrians*, in Tamil, *nīreṭṭimuttu*, commonly known to the British as 'marotty', was often used by the British as a substitute for chaulmugra. *Hydnocarpus alpina* and *Hydnocarpus wightiana*, the latter a source of chaulmugra oil, were related species. Marotty oil had a nauseous smell and an oily acrid taste. In appearance it was very similar to almond oil, though far thicker in consistency, and in chemical effects it resembled chaulmugra oil. Marotty oil was held in great repute in indigenous medicine in south India as an external treatment in leprosy and other cutaneous conditions. However, unlike *Gynocardia odorata*, the plant believed by the British to be the source of chaulmugra oil, marotty, listed by Waring under the name *Hydnocarpus inebrians*, was given only 'non-official' status in the *Pharmacopœia of India*.[45] Ainslie had already noted the importance of marotty oil in indigenous treatment of leprosy, listing it under its Tamil name as 'neeradi moottoo' in the *Materia Indica*, although he was unable to identify the actual plant from which the oil was made. By the 1860s, Waring was able to state the close botanical relationship between chaulmugra oil, believed to be *Gynocardia odorata*, and marotty, identified as *Hydnocarpus inebrians*, and the high repute in indigenous medicine of the *Hydnocarpus inebrians* seeds as a remedy for leprosy. On the basis of their close relationship, Waring recommended that marotty be given a trial as a leprosy treatment especially in the regions where chaulmugra was not procurable. Ironically, unlike *Gynocardia odorata*, which had official status in the *Pharmacopœia*, *Hydnocarpus inebrians* was not yet recognized as a remedy deserving of use in British medical practice in India and was listed as 'non-official'.[46]

The Madras Leper Hospital reintroduced chaulmugra oil as a leprosy treatment in June 1874. Writing in 1875, Surgeon W. Macrae noted that marotty oil had been used in the Madras Leper Hospital under the name of 'chaulmoogra or neeradee mootoo' many years previously. Thus, it is likely that the chaulmugra oil mentioned by van-Someren as a leprosy treatment in the first-half of the nineteenth century either referred to or was used interchangeably with marotty oil, specially since marotty was available cheaply in the Madras bazaar.[47]

It was not until the end of the nineteenth century, after the decline in favour of gurjon oil, that chaulmugra came to the attention of the medical profession in Europe. As an indication of the influence of gurjon oil, lectures on *Dipterocarpi balsamum* (*Dipterocarpus lævis*), or gurjon oil, were included in the syllabus for the course of lectures on indigenous therapies given at Madras Medical College from 1895–98, but the Hydnocarpus group of oils, which played a vital role in south Indian treatment of leprosy, were not discussed. In 1929, however, the Madras *Civil Medical Code* listed chaulmugra, still mistakenly identified as *Gynocardia odorata*, as an important palliative leprosy treatment and recommended that efforts be made to extend its use.[48] Chaulmugra, known as Hydnocarpus oil, and derived principally from the seeds of *Taraktogenous kurzii* from Assam and Burma, *Hydnocarpus wightiana* from south-west India and *Hydnocarpus anthelmintica* from Siam and Indo-China was to become the core of British treatment of leprosy in India. The Hydnocarpus oils and their esters were either taken orally, applied externally or, from the early twentieth century, injected. Hydnocarpus oil was usually injected hot, despite the difficulties of keeping the oil both hot and free of dust, particularly in out-patient leprosy clinics with minimal facilities.[49]

The Hydnocarpus oil treatments remained the most successful remedies available and the indigenous Indian treatments for leprosy became the principal form of leprosy treatment employed not only by practitioners of European medicine in India but wherever leprosy was prevalent. The International Leprosy Congress held in Cairo in 1938 declared that, world wide, 'Hydnocarpus oil and its esters administered intramuscularly, subcutaneously and intradermally, remain, so far as our present knowledge goes, the most efficacious drugs for the special treatment of leprosy.'[50] British and European recognition of chaulmugra occurred despite Home and Indian government neglect of indigenous remedies for leprosy and their preference, from the 1870s, for European- and British-developed remedies. Ultimately, the local British medical officer's practice of using the best available remedy at the time in leprosy treatment won out over any idea of the superiority of British medicine to indigenous.

Dougall's development of the gurjon oil remedy occurred when, increasingly, a distinction was being made between indigenous and European medicine and between indigenous and European conceptualizations of leprosy, with European medical knowledge being privileged over indigenous, particularly at the British and Indian government level. In a letter to the *Indian Medical Gazette* of June

1877, Dr P.C. Banerjea, who had been moderately successful in treating leprosy with chaulmugra oil, noted with chagrin that since the British implemented treatment with gurjon oil, described as 'a new medicine' and 'a foreign drug', all other remedies for leprosy, especially those traditionally used in India, had been neglected.[51] Certainly, government endorsement of trials for gurjon oil, and refusal to endorse trials of chaulmugra oil, indicated a neglect of indigenous remedies at higher Government of India and Home government level. However, British medical officers directly involved in the care of leprosy sufferers did continue to use indigenous remedies, particularly chaulmugra oil. It is likely that Banerjea's letter refers to changes in indigenous medical practice as much as to a decrease in British attention to indigenous remedies.

In the area of leprosy treatment at least, British adoption of indigenous remedies was not a conservative process but entailed change both in the range of conditions treated by the adopted indigenous remedies and in the emphasis placed on the correct combination of indigenous substances and their integration with a diet regimen. Dougall adopted a remedy already used in the British medical system as a substitute for an imported British remedy and extended its application in the British treatment context to include leprosy. Similarly, when Waring added *Hydrocotyle asiatica* into the *Pharmacopœia of India*, he did not simply record the existing indigenous uses of the substance, but gave recognition to the extension of the substance's therapeutic range in British medicine to include leprosy treatment.[52]

Typically, British use of indigenous treatments for leprosy involved isolating one substance, such as chaulmugra or gurjon oil, from the wider indigenous treatment context, transferring it into a British regimen of diet and hygiene and employing it as the sole medicinal treatment, a 'specific' cure for leprosy. Although British and indigenous medical systems were broadly comparable in their understanding of diet and hygiene as important aspects of medical treatment, and an 'improved' diet was often prescribed for leprosy patients, the British placed far more emphasis on the power of the medicinal agent to effect cure. The hope of a 'magic bullet' for leprosy which only became a medical reality with the development of Dapsone in the 1940s, was already, in the nineteenth century, part of British medical conceptualization of effective leprosy treatment. In British trials of gurjon oil for example, according to the principles of experimental method, the diet was left unchanged in order to better perceive the medicinal effect of the drug. Similarly, where chaulmugra was employed by British medical officers there was no mention of its being supported with neem oil or any other substance, as occurred in indigenous treatment regimens. Further, Mouat's recommendation in 1854 that to increase chaulmugra's efficacy, the patient's diet be carefully regulated according to indigenous treatment methods went unheeded. The description in Maclean's *Glossary*, published in 1893, of the necessity to combine dosage of chaulmugra oil with neem or some similar oil[53] is significant for its advocacy of indigenous treatment methods. Maclean's recommendations for the use of chaulmugra oil in treatment suggests not only that British understanding and knowledge of indigenous medicine had vastly increased by the close of the nineteenth century, but that at the presidency level, retaining indigenous combinations of remedies was being recognized as of medical value. Arnold argues convincingly that 'appropriation, subordination and denigration were the processes by which European medicine marked its conquest over indigenous medicine' at least rhetorically, even if not so completely in practice.[54] By the end of the nineteenth century, in the area of leprosy treatment respect for indigenous remedies and treatment methods, rather than denigration, was regaining ground. British medicine was sounding the retreat in the treatment of leprosy.

Indigenous borrowing from British medicine

Although it has become customary to focus on colonial appropriation of indigenous remedies, practitioners of Indian medicine were, in general, willing to employ British remedies if they appeared more effective than the available indigenous treatments. Ironically, bleeding, which was considered by the British to be too strenuous a treatment for Indian patients, was, by the early eighteenth century, frequently employed in Ayurvedic medicine. A medical officer reporting on Dindigul in the early nineteenth century noted that a few European medicines were requested and when available, used by the Indian practitioners within the context of their Siddha approach to the treatment of disease. Although Siddha medicine did not hold the political and administrative power in British India to enable the institutionalization of such borrowing, clearly, Indian medical practitioners as well as British were willing to extract medical remedies from a foreign culture to serve their own medical needs.

Despite the British rhetoric of medical supremacy, it is unlikely that practitioners of indigenous medicine considered that such borrowing actually compromised their medical authority or the integrity of the indigenous medical tradition. In his 1917 monograph on the role of chaulmugra oil in Ayurvedic medicine, J.C. Ghose recommended the internal, external or hypodermic administration of the European remedies, Fowler's solution or Donovan's solution in conjunction with chaulmugra oil as part of the Ayurvedic treatment of leprosy.[55] Just as the British incorporated specific indigenous substances into their own treatment regimens, British treatments were included within existing combinations of indigenous remedies. Only in the early twentieth century was British rhetoric addressed head on, and restoration and promotion of indigenous medicine as equal, if not superior, to British became an aspect of nascent Indian nationalism. In Madras renewed Indian interest in indigenous medicine was marked by the enquiry into indigenous drugs established in 1917 under the authority of M.C. Koman, honorary physician to the General Hospital, whose first report was submitted in December 1918.[56]

Conditions facilitating exchange between British and indigenous medicine in the nineteenth century were always, however, delicately balanced. In circumstances of overt government intervention in indigenous treatment methods, where Indian religious and cultural sensibilities were subordinated to the claims of British medicine for superior effectiveness, indigenous openness to the use of and treatment with European medicine could swiftly be replaced by resistance and opposition by both medical practitioners and patients. The abolition of the indigenous method of smallpox variolation and implementation of vaccination as the only form of smallpox treatment by the British in 1865 is one case in which such a resistance occurred.[57] In the context of leprosy treatment the dynamic of resistance and compliance was played out first at the local level.

Patient resistance

Resistance among in-patients of all races in British administered leprosy hospitals was a significant factor affecting treatment with both indigenous and British remedies. For the leprosy patient, British concerns to impress on the Indian populace the superiority of British medicine meant little. Not all those with leprosy in the asylums were entirely compliant with medical efforts to treat their disease or to test possible cures. Dougall had the good fortune, from his medical colleagues' point of view, to be in charge of a penal colony with an unusually stable, permanent and co-operative leprous population on whom to test his cure. As Cornish commented in his 1880 report: 'There is no law under which lepers can be restrained in an asylum. Many come for a time and give a trial to medical treatment, but getting tired of restraint leave the hospital, returning ultimately to die.'[58] Even after the passing of the 1898 Lepers' Act, which legislated for confinement of vagrant leprosy sufferers, without co-operation, restraint was virtually impossible.

Hospitalization for long-term illness was not customary among Indians, though in the Madras presidency from the early years of the hospital and dispensary system, assistance was sought in British hospitals and dispensaries in cases of surgical need or when severe accidents had occurred. In cases of illness, the officer reporting on medicine in Nellore in the early nineteenth century noted:

> The natives... frequently resort to a change of climate on a long continuance of any disease, and they often on such occasions undertake a pilgrimage to some sacred pagoda or temple, but their birth place or the residence of friends is more frequently selected.[59]

Asylums tended to be frequented only by those leprosy sufferers who were without homes or friends or whose family was too poor to support them. Hindus and Moslems of the higher classes, for social and religious reasons, or because they were more able to rely on their better-off family and friends, tended to avoid medical assistance and residence at European hospitals and dispensaries.[60] Typically, leprosy sufferers resided temporarily in hospitals because of destitution rather than in the expectation that, after a certain course of treatment, they would be cured. There did not exist the cultural assumption that severe illness required residence in hospital.

Similarly, despite the reports by British medical officers of a favourable Indian reception of civil hospitals and dispensaries, there was still, in the late nineteenth century, a widespread and deep cultural reluctance to seeking European remedies for illness. European medicine was often a last resort. The report on the civil dispensary at Bellary by the garrison surgeon was typical. The surgeon noted with frustration that most cases presenting at the dispensary were of a severe form and rendered greatly more complicated and distressing by previous 'maltreatment': 'for here, as elsewhere in India, Natives are so confirmed – with regard to the employment of country remedies, that they seldom apply for European advice until having been ineffectively treated by Native doctors.'[61]

Even when, with Dalton's cure and the gurjon oil remedy, leprosy sufferers did become drawn to asylums and hospitals in the hope of cure, there were many factors mitigating against sustained, long-term co-operation with treatments offered by the British. In the Madras presidency, many medical officers attempting to test the gurjon oil remedy complained at the resistance of leprosy sufferers to treatment. Medical officers often attributed this lack of enthusiasm by the leprosy sufferers to stupidity or laziness. Surgeon-Major W. Macrae, reporting on the trials of gurjon oil at the Triplicane Dispensary in 1875, commented with some irritation, that 'lepers are far too indolent to take the trouble of rubbing themselves for long periods every day, and unless there is someone present to see that they do it, the work is very slipshod'. More sympathetic officers such as Day, who had a greater knowledge of leprosy, recognized such a response as, at least in part, an effect of the disease:

> The nature of this disease appears to render these unfortunate beings extremely obstinate; they become morose, inclined to brood in solitude, drink spirits, eat opium, smoke ganjah; and in fact do anything, to wean them from a recollection of the past, a view of the present, or a thought of the wretched earthly future before them, terminating only with their existence.

The high degree of regimentation and benign compulsion employed in Dougall's treatment regimen, was, in part, intended to combat such depression.[62]

It was a source of great frustration to medical officers that patients not only resisted treatment but some ran away from asylums, foiling their efforts to provide care and test leprosy treatments. In civil hospitals and dispensaries to which inmates other than leprosy sufferers were admitted, sometimes fear of infection with other diseases caused the leprous patients to leave. At the Chingleput Dispensary, the admission of a cholera patient provoked the exodus of a large number of leprosy sufferers who had come for gurjon oil treatment. Further, although many asylums made considerable efforts to make conditions comfortable for leprosy sufferers, the inevitable tedium of institutional life and the often inadequate diet and facilities in institutions not specifically appointed for leprosy care, like most civil hospitals, provided leprosy sufferers with little incentive to remain.[63]

For those of the poorer classes who made up the majority of residents in the leprosy hospitals, even the best conditions could not always compensate for the loss of freedom, the relatively good living and the interesting life which could be gained by begging. Surgeon-Major W.H. Roberts reported his frustration with attempts to give gurjon oil a fair trial at the Calicut Leper Hospital under very different conditions to those experienced by Dougall at the Port Blair penal colony:

> [At Port Blair] the leper convict had no aim or object in life till fired with the hope this treatment afforded of being alleviated or rid of his disease. Possessed of this healthy stimulus, he worked away at his own cure with a faith and a will that invariably bear fruit, good or bad, as the case may be. Here the mendicant vagrant leper never for a moment forgets that he is a free man, free to come and go as he pleases, and this liberty of action he asserts on every, even the slightest, provocation. As feasts, fasts, and fairs come round, he must needs attend them to indulge in his begging proclivities, and if these do not suit his whim, he simply walks away. Thus treatment is interrupted, the wish to get cured is not so ardent, or at any rate is not controlled by that *will* which is an important factor in this treatment.[64]

Few medical officers in search of a cure were willing to accept the preference of many leprosy sufferers to live independently of the asylum by begging and the consequent need to preserve some obvious signs of disease. Roberts' comments reflected not only a belief in the superiority of British medicine over indigenous but in the greater authority of the doctor in the doctor/patient relationship. However, not all medical officers were as blind as Roberts was to the autonomy of the leprosy sufferer. Some recognized that the desire to attend fairs and to beg was not an irresponsible flouting of British medical authority, and by implication, of British imperial authority, but an essential part of the leprosy sufferer's life and gaining of livelihood. The available treatments which, when successful, markedly reduced external signs of leprosy, such as roughness and irritation of the skin, sores and ulceration, were simply impractical for those who lived by begging. Writing in 1875 on the response of leprosy sufferers to gurjon oil treatment, the zillah surgeon at Berhampore counted among his most reluctant patients: 'professional mendicants; [who] if they allow themselves to be cured would be deprived of an easy and certain livelihood'. The situation had altered little from the early nineteenth century. Dalton believed that those leprosy sufferers who left the Madras Leper Hospital before he considered them to be cured were, 'beggars who after they had lost the great irritability attending the disease preferred to remain ulcerated in order to gain commiseration from the public and proportionate alms'.[65]

Financial stress and responsibilities at home, particularly among those not dependent on begging for a livelihood, rather than a preference for non-institutional life, prevented other, perhaps more willing patients from completing long-term treatment. The surgeon at Chingleput Dispensary reported in 1875 that among the leprosy sufferers who sought admission for gurjon oil treatment, most had spent their time in hospital at two or three different periods, often expressing a wish to leave and return home briefly before continuing treatment. The assistant apothecary at Bimlipatam Civil Dispensary reported that attempts to give gurjon oil a long-term trial were impeded because villagers, living close to the hospital but admitted as in-patients, were often forced to discontinue treatment since 'they had no means to support themselves', while their village work was left neglected. For villagers who had to travel considerable distances to reach the dispensary, long-term regular treatment of doubtful benefit would have been impossible at busy times such as the harvest season. Often, such reluctance to pursue treatment was interpreted by the British as a lack of patience and resolve, typical of the Indian patient. The medical officer reporting on the 'state of medical science' in the Coorg, commented that many Indians sought assistance from British medical practitioners; 'like all natives however, they want patience to submit to any lengthened course of treatment, and generally return to their homes if a cure is not effected in a few days'.[66] However, not all medical officers lacked insight into the practical realities of poverty endured by those who sought leprosy treatment at British institutions. By some at least, it was understood and seen as quite reasonable that many with leprosy placed the need to sustain their own and their families' livelihood ahead of the uncertainty of hospital cure.

That other leprosy sufferers were willing to receive food and shelter in government asylums but not necessarily to submit to treatment, suggests that government was unable to convince their patients that the treatments they offered were truly powerful and effective. Many were too disillusioned by the repeated failures of cures to submit readily to treatments which were often not simply unpleasant but, at times, acutely painful. Van-Someren reported that during the trial of Beauperthuy's remedy at the Madras Leper Hospital, patients shrank from the application of cashew nut oil because of the pain it caused. In one case he recorded: 'the patient refused to submit to a second application, and in several instances among the East Indians [Eurasians] I was obliged to hold expulsion from hospital *in terrorem* over them before they would submit to treatment.' Despite Dougall's glowing reports from Port Blair, there were also reports from the Madras presidency hospitals and dispensaries that some leprosy patients disliked the gurjon oil treatment and either refused to participate or left the asylum to avoid it.[67] It is likely that either direct experience or the reputation of Beauperthuy's treatment, trialled in the Madras presidency during 1873, contributed to mistrust among some patients of further remedies for leprosy presented by the British.

Notably, although those with leprosy flocked to the asylums on the strength of the rumoured curative power of British-discovered cures, once at the asylum treatments from the indigenous tradition were most readily accepted. At Palliport, the preference of many patients for fumigation was possibly due to its familiarity as a technique for leprosy treatment in Siddha and Ayurvedic medicine as well as to the lack of effort required in comparison with the gurjon oil treatment. Even more striking was the preference among patients at both the Madras Leper Hospital and Palliport Lazaretto for chaulmugra oil. Surgeon Thompson, in his annual report on the Madras Leper Hospital for the official year 1876–77, noted that 'on the whole the patients prefer the chaulmoogra oil' to either gurjon oil or a combination of ordinary turpentine mixed with coconut oil which was very similar in its effects to the gurjon oil treatment. Thompson described the patients' preference in terms of their experience of the treatment, chaulmugra being 'milder in its action' and 'not disagreeable to the feelings after it has been well rubbed in' and noted that the patients 'think too, it keeps the skin softer and more pliable'. At Palliport, the question of efficacy was specifically raised in the report of the same year. Surgeon-Major A. Gamack observed that of fumigation, gurjon, cashew nut and chaulmugra oil treatments, chaulmugra was the most popular with the patients 'and is, in fact, the only treatment in which they appear to have any faith'.[68] Medical government authorities at home strongly endorsed European- and British-developed remedies as specifics for leprosy. However, in the leprosy hospitals, local British medical officers continued to use indigenous remedies, and the leprosy sufferers themselves tended to regard them as more efficacious than introduced treatments.

Resistance by leprosy patients to treatment offered in nineteenth-century government institutions also proceeded from the understanding, among Indian leprosy sufferers at least, that the disease required spiritual as well as physical healing. Reflecting both the religious character of indigenous medicine and the belief that karmic factors caused leprosy, the most important indigenous remedies for leprosy had religious associations. The ancient Burmese myth of the cure of Rama and Priya with chaulmugra became linked in popular legend with the god Rāma of the Hindu *Rāmāyaṇa*.[69] Neem also had ancient religious links with the healing of leprosy, being specifically linked to leprosy in the village context and in temples through *nāga* (snake) worship. It was commonly believed in south India that anyone who killed a snake, particularly a cobra, even if accidentally, would be punished in this or a later life, with either barrenness, ophthalmia or leprosy. Images of *nāga* were installed in temples and other sites on a platform prepared in the shade of a pipal or margosa tree, a relic of ancient tree and serpent worship. Those with leprosy, thus, often visited a serpent shrine to appease the serpent they may have angered in this or a previous life. Further, the same neem plant which was used in indigenous leprosy treatment was closely identified with village goddesses such as Kāliamma in Tamil areas and Pallalama in Telugu villages. These goddesses were believed to be both the source of epidemic disease, particularly cholera, and the means to protect the village from its ravages. In most Tamil villages, Mariamma was worshipped specifically as the goddess of smallpox. Neem was used to decorate the brass or earthenware pots which represented the goddess in village festivities and to garland the cattle sacrificed to appease her.[70]

Such a concept was not unique to India but also had a long European tradition. For some medical officers at least, although religion was not publicly employed in leprosy cure, there was a religious dimension to healing leprosy sufferers. Dalton, in 1813, described leprosy as an 'Evil' and gave 'thanks to Almighty God for having made me an humble instrument if not in curing them in affording the greatest relief'.[71] However, with the development of a more scientific approach to medicine and the secularization of medical culture, the religious aspect of leprosy cure declined in importance during the nineteenth century.

From the Hindu point of view, leprosy, though understood as having physical causes, in the nineteenth century was still firmly believed to be the result of 'sinful taint', the mark of bad action in a previous life. Consequently, physical healing without reference to the supernatural could not fully cure the disease. It is likely that much of the 'faith' in chaulmugra oil among leprosy sufferers at Palliport was due to its supernatural associations. However, even at Palliport, where familiar indigenous treatment techniques such as fumigation and chaulmugra oil were accepted by some patients, others avoided treatment. It is likely that even traditional therapeutic methods, stripped of the necessary ritual and presented in the alien healing context of a government asylum, were perceived by some Indian patients as essentially meaningless and thus ineffective. In 1876, the civil surgeon described, in his report on Palliport Lazaretto, the leprosy patients' avoidance of treatment and flight from the asylum as 'a deplorable account of the unwillingness of these poor people to do anything for their improvement, most probably because they think it is vain trouble'. Revealing a typical late-nineteenth-century British contempt for indigenous belief in the efficacy of religious healing, he continued:

> I have no doubt, however, that if any painted conjurer, smeared with ashes and filth, was to appear among them with incantations, & c., and offer to cure, they would eagerly crowd round him and submit to very cruel burnings and scarifyings, and probably even part with any small store of pice which they may have collected.[72]

The Indian response to purely secular treatments was not unlike the mid-nineteenth-century Indian resistance to the European 'godless vaccination' against smallpox which neglected traditional Indian placation of the smallpox goddess, worshipped as Mariamma in the south and Sītala in the north. For some Hindus it made sense to mistrust exclusively secular therapies for leprosy and instead rely upon divine intervention for recovery. It is likely that the failure of traditional Indian religions to deliver cure contributed to the conversion of leprosy sufferers to Christianity. The rate of conversion was so great that Jonathan Hutchinson, known for his theory of a relationship between leprosy and eating fish, suggested that Christianity, and in particular Roman Catholicism, which encouraged fish eating on specific days, was a contributing cause of the disease. In Madras, as throughout India, the actual incidence of leprosy among Indian Christians, except for those who were converted to Christianity in asylums was, however, well below the average.[73]

Recognizing the centrality of reparation to the Indian understanding of leprosy cure, the High Priest of the Baidyanath healing temple in Bengal argued strongly during the late-nineteenth-century debates on the proposed Lepers Act that the Government of India should recognize the significance of temples and shrines throughout India in the life of the leprosy sufferer. He proposed that within the terms of the government's efforts to reduce contagion by segregation, the role of temples in feeding, sheltering and offering a chance for supernatural intervention to those with leprosy should be formalized, with the government financing separate accommodation for leprosy sufferers at popular temple complexes. Those leprosy sufferers confined in government or mission asylums, which did not cater to their need for 'Divine interference', should be 'permitted to attend the temple at specific times, and under proper restriction'. The high priest was not alone in his concern for the spiritual needs of the leprosy sufferer. The editor of the Calcutta paper, the *Indian Mirror*, endorsed his recommendations and argued that 'due respect should be paid to the religious beliefs which leads the Hindu lepers to spend their last days in sacred places of Hindu pilgrimage'.[74] The relatively high degree of co-operation with treatment seen among those at the Madras Leper Hospital in comparison with those at Palliport was very likely to have been due, in part, to the wide provision of religious rites at the Madras Hospital. At Palliport, Roman Catholic religion and ritual was available, and many of the patients were Roman Catholic. However, from the evidence of the civil surgeon, the combination of ritual and healing power available in the indigenous tradition still seemed most efficacious to the leprosy sufferers and was neither endorsed by the British at Palliport, nor on offer.[75]

Patient resistance to treatment was not, as some British medical officers suggested, a consequence of the indolence and irresponsibility of, particularly Indian, leprosy sufferers. Rather, it indicated that the social and economic concerns of leprosy sufferers and the indigenous perception of leprosy as, in part, a spiritual condition could not be adequately addressed in the British treatment context. It was not until the introduction in the 1950s of Dapsone, an anti-bacterial sulphone drug, effective against the leprosy bacillus, that leprosy sufferers in India could begin to hope for a complete medical cure (though the swift development of Dapsone-resistant leprosy strains made this only briefly possible) and the need for supernatural intervention became less pressing.

As Arnold argues, the very limited dominance of European over indigenous medicine by the twentieth century was achieved in part by the hegemonic relationship between British and Indians. This relationship was particularly evident in the co-operation and participation in promotion of British medicine as superior to indigenous by Indian doctors trained in British medicine and employed in the IMS. At the same time that, according to Bala, the superiority of English drugs irrevocably split British and Indian medicine, Madras medical authorities were, however, continuing to endorse the employment of unmanufactured indigenous medicines in the practice of British medicine. W.R. Cornish, then surgeon-general with the Madras government, complained to the chief secretary in 1880 that Indian medical subordinates preferred using imported British remedies to indigenous treatments included in the Indian pharmacopoeia, and attributed this reluctance, in part, to a lack of training in the use of indigenous remedies.[76] In the area of leprosy treatment, the turn of the century brought not British medical dominance but British and international championing of chaulmugra oil, the principal indigenous Indian remedy for leprosy. The divergence of European from indigenous medicine and the triumph of the Western medical drug culture was far less complete than Bala suggests. The persistence of indigenous forms of leprosy treatment into the twentieth century and the resistance of leprosy patients to both British and indigenous treatments offered by the British lends further weight to arguments for the weakness of colonial medical intervention. The relationship of British medicine to indigenous medical systems cannot be described as one of unfettered British superiority.

5
Leprosy Research and the Development of Colonial Medical Science

In the first half of the nineteenth century, investigation and research into the disease of leprosy and the search for leprosy remedies in the Madras presidency were pursued primarily by interested individual medical officers with some regulation by the Madras Medical Board and presidency government. Government attention was, at most, spasmodic, since care of leprosy sufferers and the incorporation of leprosy remedies into the British medical system in India were not a government priority. From the 1860s until the close of the 1870s, during the period of expansion and consolidation of crown rule in India, the attention of British government and Home medical authorities was drawn to leprosy in the empire. The investigation of leprosy in India developed in the wider context of British concern for the welfare of the empire and the culture of scientific enquiry burgeoning in Europe.

Introduction

Basalla's model of Western science

In 1967, George Basalla published an article on the spread of Western science from Europe into the colonial world. Though not uncritically received, Basalla's article has considerably influenced understanding of the spread of science into the non-European world, including India. Applying the article to the Indian context, Arnold justly criticized Basalla for being 'diffusionist, rather than interactive or dialectical' and thus neglecting the importance of 'local constraints and impulses' and 'political and professional influences' in shaping the development of European science in India.[1] Even so, Basalla's model, though not specifically referring to medical science, provides a broad typology in which to place the history of European medicine in India. Basalla describes the spread of Western science to the non-Western world as occurring in three phases. In the first phase, the non-scientific society provided a source of information and ideas for European science. In the second phase, a period of 'colonial science' emerged during which the European scientific culture developing in the colonies remained dependent on European scientific culture but not servile to it. In the third phase, the transplantation of Western science into the colonial context was completed, characterized by an effort to define and assert a scientific identity independent of Europe.

Like Basalla's model of colonial science which suggests an overlap between phases,[2] in India the first phase, evident in early European investigation of indigenous medicine, persisted to the close of the nineteenth century. The second phase gathered momentum in south India from the mid-nineteenth century with the establishment of crown rule, the development of an increasingly scientific medical culture in Europe and a related scientific medical culture in India. However, while recognizing an overlap of phases, Basalla's model neglected the interactive relationship between European and indigenous medicine so evident in the Indian context. In India, through the dynamic of medical interaction between European and indigenous culture, the first phase contributed to and co-existed with the development of the second phase of colonial medicine as an entity somewhat independent and separate from medicine in Europe.[3]

The study of leprosy investigation suggests that the seeds of a colonial medical culture in India are to be found embedded in phase one in the early nineteenth century. Not only did phase one of Basalla's model persist throughout the nineteenth century, but from the earliest decades, a form of colonial medical science co-existed with an understanding of Indian medicine as a source for British medicine. It is beyond the scope of this book to examine the significance of leprosy with regard to the emergence in India during the early twentieth century of the third phase of colonial science and the attendant revival of indigenous medical systems linked to the rise of nationalism.

Impediments to leprosy research throughout the nineteenth century

Throughout the nineteenth century, research into leprosy and the discovery of a cure for the disease were low priorities for the East India Company (EIC) and later crown authorities, not only in the Madras presidency but also throughout India. Leprosy was principally to be found among Indians and Eurasians who, as a whole, received little British medical attention compared to the European military and civilian population. Those Indians with leprosy who came to the attention of the Indian Medical Service were usually male, poor to the point of destitution and homeless. Those of the middle classes afflicted with leprosy were rarely noticed, especially women and girls who were kept at home and thus were more able to hide the disease from public attention. As a consequence, leprosy, identified primarily by the British with Indian men of the poorest and least powerful classes, was not perceived as a disease which posed any significant threat to the trading power of the EIC and, even after the events of 1857, to the strength of the European army or the British image in India. Further, unlike typhoid or plague, when leprosy did occur among Europeans it affected few and the sufferer could be sent home to England,[4] thus sparing embarrassment to the ruling race.

It was not only medical research into leprosy which was neglected. Throughout the nineteenth century two factors mitigated against any form of medical research by the Indian Medical Service (IMS). The first was the lack of medical qualifications among IMS recruits, particularly in the first half of the nineteenth century, combined with a deeply conservative approach to medicine and medical treatments. The second was a reluctance to initiate change. It is likely that the increased attention to medical research in the second half of the century was, in part, a result of improvement in the educational quality of IMS recruits, although even then, only a little over 5 per cent of IMS men recruited between 1857 and 1887 held medical degrees. Most IMS officers who made a significant contribution to the understanding and treatment of leprosy from the 1860s onwards held medical qualifications. Van-Someren, Carter and Dougall all held an MD as did G.C. Roy, one of the very few Indian recruits to the IMS during the nineteenth century. The IMS was opened to Indians only in 1855 and entry examinations were held in England. For those IMS officers who were interested in scientific medical investigation, chronic lack of resources and under-funding were a constant source of frustration throughout the nineteenth century.[5]

In addition to the lack of incentive from government for investigation of leprosy by the medical service, there were positive deterrents to its study which significantly hindered medical understanding of leprosy and the search for its cure. The loathsomeness of the disease, its intractability in the face of medical intervention and the fact it would bring neither honour nor wealth to the investigator, it being primarily observed among the poor Indian population, strongly discouraged the investigation of leprosy by the British medical profession in India. For those who might have considered studying leprosy, it was often logistically too difficult to research a disease accessible to the medical practitioner almost exclusively within the vagrant Indian population.[6]

Advances in understanding and treatment of the disease in India were usually made by those who also had an altruistic concern for the leprosy sufferer's welfare. Medical officers appointed to leprosy hospitals and who thus had every opportunity to study the disease, were reluctant to court even proximity to, let alone close scrutiny of their charges unless they felt a distinct affinity with the leprous poor. Dougall, who arranged for leprous convicts at Port Blair to be photographed at his house so that he could document their improvement in health, was the exception rather than the rule. Apothecaries at the Madras Leper Hospital bitterly complained about the quarters made available to them at the leprosy hospital from the mid-1870s. The apothecary's quarters were surrounded on three sides by wards, the closest, which accommodated 20 leprosy sufferers, being only one yard from the apothecary's quarters. Cook, the surgeon in charge, commented in 1888 that, even leaving aside the possibility that leprosy was contagious, he considered it 'a hardship, if not inhuman, to force any man to be within constant sight of about 40 lepers while sitting in his own private quarters'. G. Bidie, the surgeon-general, endorsed Cook's view, adding that 'it is not right to compel any healthy person to live so near the lepers'. Not only was the sight of lepers 'at all stages of the disease' offensive, but 'at certain times of the year the smell of the sores invades the premises'.[7]

Even though leprosy research attracted little official kudos throughout the nineteenth century, the discovery of a cure, especially when pursued apparently altruistically and without expectation of financial reward, did carry the stamp of professional respectability and even merit. In 1874, Dougall's 'disinterested labours' on behalf of the leprosy sufferers of Port Blair without thought for financial gain, were praised by the chief commissioner of the Andaman and Nicobar Islands as evidence of his 'professional character'. By the 1870s, when leprosy had come to be regarded by both the medical profession in Britain and the British government as a matter of colonial responsibility, the discoverer of an effective leprosy cure could have expected payment from the British government in return for allowing its use throughout the colonies.[8]

Leprosy research 1800–60

Independent treatment initiatives and peer supervision

In the first half of the nineteenth century, medical care of leprosy sufferers and investigation of remedies for leprosy were principally local matters, involving a small section of the medical service, the Madras Medical Board and the Madras government. A definitive understanding of leprosy based on the description of external symptoms was not available until the researches of Daniellsen and Boeck were published in the 1840s. Coherent attempts by the British medical profession in England and the Indian and colonial governments to investigate the character and etiology of leprosy did not develop in south India until the second half of the nineteenth century. In the first half of the century, medical research in relation to leprosy principally consisted of individual efforts to discover, analyse and test possible remedies for the disease from both the British and indigenous traditions. The principles of empirical scientific method and tools of chemical analysis were employed. From the early nineteenth century, these principles gradually displaced the remnants of humoral medicine and laid the groundwork for the modern scientific medical understanding based in observation, experimentation and inference rather than theoretical speculation.

Research and scientific investigation of leprosy and of possible cures for the disease were not specifically encouraged by the Madras presidency government, and IMS and Indian government interest was non-existent. However, the subjection of proposed leprosy treatments to 'thorough and long-continued trial', and keeping the Madras government informed of treatments through the medical board and annual reports, were well-established practices by the 1850s. Dalton brought his mercury treatment to the attention of the Madras government in 1811, requesting that the governor-in-council order the 'Medical Gentlemen' of the presidency to inspect patients who appeared to be cured. Dalton offered to make his remedy available for trial elsewhere if deemed appropriate and to promulgate the recipe if the remedy proved effective.[9]

The committee overseeing the trial found the cure to be more a threat to the leprosy sufferer's life than a remedy for the disease and by July 1813 the Madras government was convinced that Dalton's cure was ineffective.[10] There was no widespread testing of the cure either independently by the medical profession or with Madras or Indian government endorsement as occurred in the second part of the nineteenth century with the treatments proposed by Beauperthuy and Dougall, which, on the basis of initial local trials, were no more effective than Dalton's cure. It was not the relative merit of the cures which had changed: Beauperthuy's use of mercury was reckoned in the 1870s to be as dangerous to the leprosy sufferer as Dalton's. Greater British government attention to leprosy in India from the 1860s made both the British and Indian governments more interested in investigating alleged leprosy cures by the 1870s than they had been in the early nineteenth century.

Scientific exchange

Although there was little Indian government interest in investigation of leprosy remedies in the first half of the nineteenth century, local medical officers in Madras did not work in isolation. By the 1850s there was considerable exchange in medical information concerning indigenous remedies both among British colonies and between the British and their European colonial rivals. In particular, the British medical officers were aware of the research and investigations pursued by the French. In the early nineteenth century, the French were further advanced in the professional pursuit of 'scientific medicine' than the British,[11] and looked beyond their Indian possessions to further their knowledge of possible indigenous medicines for intractable diseases such as leprosy.

The French and later British investigation of *Hydrocotyle asiatica* was a case in point. In the French south-Indian territory of Pondicherry, Jules Lépine performed an exhaustive analysis of *Hydrocotyle asiatica* after receiving favourable reports from Dr Boileau of Mauritius, who had employed it in treating his own leprosy. The analysis was subsequently published in Pondicherry's *Le Moniteur Officiel. Hydrocotyle asiatica* was not employed in Siddha medicine as a remedy for leprosy, although some practitioners recommended its use in the form of a paste in the treatment of venereal ulcers. Rather, it was one of the plants employed in Siddha *kāyakaṟpam*, or rejuvenative treatment, believed able to cure degenerative diseases and to ensure long life.[12] The French tested the therapeutic uses of the plant in leprous, venereal and other conditions with beneficial results in the majority of cases. Seven editions of *Le Moniteur Officiel* reporting on the *Hydrocotyle asiatica* and its uses were forwarded to the Madras Drug Committee.

Hydrocotyle asiatica was then known to the British as an indigenous treatment for bowel complaints and fever and, in the treatment of infants, as a diuretic and alterative. Ainslie recorded that, on the Coromandel Coast, the leaves were externally applied in indigenous medicine as an anti-inflammatory for soft-tissue damage and bruises. *Hydrocotyle asiatica* was trialled by the British at the Madras Leper Hospital and, later, the Native Infirmary to compare the results with those of Lépine and Dr Boileau in their Pondicherry trials.[13] On the basis of their own observation of the trials at the Madras Leper Hospital, the members of the medical board found that although the plant was a useful tonic and alterative, it did not remove the 'constitutional taint' of leprosy and thus could not be considered a cure, a view held also by A. Hunter, assistant surgeon in charge of the Native Infirmary, Madras, who reported at length on the remedy. Reflecting inter-colonial medical exchange, Surgeon Hunter's 1855 report was forwarded to the British colonial government at the Cape of Good Hope in response to their request for information on the use of *Hydrocotyle asiatica* in the treatment of leprosy. By 1867, no doubt in part as a consequence of the trials of *Hydrocotyle asiatica* in Pondicherry and the Madras Presidency, the *Hydrocotyle asiatica* was given 'official' ranking in Waring's *Pharmacopœia* as a treatment, both internal and external, for leprosy, syphilis, other skin diseases and ulceration. Even so, Waring was careful to note, partly on the basis of Hunter's report on the Madras trials, that *Hydrocotyle asiatica* was not proven to be a specific for leprosy.[14]

The extent of specific scientific experimentation on the *Hydrocotyle asiatica* published in *Le Moniteur Officiel* of Pondicherry and reported by Hunter is some indication of the degree of British and French interest in scientific medical experimentation in British India in the early nineteenth century. It also reveals the paucity of research resources available to the British in comparison with the French.

Jules Lépine subjected the plant to 39 experiments including the 'minute quantitative vegetable analysis' in an effort to discover its active constituents. The British in Madras were, however, poorly equipped for such research. The medical officers had fewer facilities than the French for the preparation of remedies and were not able to test the syrup or extract of the *Hydrocotyle asiatica* employed by the French. Some effort was made by the British to improve the facilities for experimentation, a 'large chemical apparatus' being manufactured in Madras so that water- and alcohol-based formulations of *Hydrocotyle asiatica* could be prepared and their therapeutic action tested on patients.[15] Even so, lack of resources for research was to remain a chronic problem for the IMS until the development of 'tropical medicine' as a recognized discipline at the turn of the century.[16]

Leprosy research in the 1860s and 1870s

European influence and the consolidation of scientific medical culture in Madras

During the 1860s the local autonomy of research into leprosy treatments in Madras altered as Home government and medical authorities began to exert an influence. Medical science at the centre of empire began to intervene directly in the development of medical science at the colonial periphery. Several factors contributed to the increased medical service and government attention to the investigation of leprosy and its treatment. The growth of scientific medical interest in leprosy in Europe encouraged research and publication on leprosy by the medical profession in India. In India, the investigation of leprosy by the medical profession was stimulated by, and contributed to, the growth of an Indian scientific medical culture. In the 1860s, the British government, guided by the Royal College of Physicians, took a direct interest in medical enquiry into leprosy, and in the 1870s, the discovery of alleged cures for leprosy further stimulated government attention. From the mid-nineteenth-century, the medical investigation and treatment of leprosy in India was thus directed increasingly by the medical profession and the government in Britain rather than by local and presidency medical and government authorities. Even so, the intervention of British governmental and medical authorities in the medical approach to leprosy in Madras, and the attendant proliferation of an investigative and regulative medical bureaucracy, did not entail the complete loss of control of medical care for leprosy sufferers at the local level. Despite government promotion of British-discovered cures, employment of indigenous remedies remained part of local British medical practice. The British medical response to leprosy is significant not only for the relationship between indigenous and British medicine at the precise level of medical treatment, but also for what it reveals of the limitations of European scientific medicine as an agent of colonial dominance.

The emergence of British belief in the inferiority of indigenous medicine to British was only one aspect of the changes in medical culture in India. During the 1860s members of the Indian Medical Service developed a more coherent methodological approach to the investigation of leprosy in response to British government initiatives in systematic investigation of disease. In the 1860s there was, at both Indian and British government levels, a widening of British medical interest in India which placed the health of the European army and civilian population into the broader context of Indian health conditions. In addition to increased anxiety to preserve the strength of the army in India after the mutiny, the transfer of British responsibility from the EIC to the crown in 1858 contributed to the growth in England of a new sense of duty to sustain and develop not only the wealth, but the welfare of India. The Government of India came under pressure from the secretary of state for India and the British parliament to initiate enquiries into disease among Indians as well as Europeans, and to take steps towards the introduction in India of sanitary reform and medical advances already occurring in Britain.[17] In addition, during the 1860s there was a movement in England away from individuals working for reform on private initiative and towards a more coherent, government-directed response to sanitary and medical problems based on the collection and dissemination of information on a large scale.[18] This shift resulted in the increasing bureaucratization of leprosy investigation in India during the second half of the nineteenth century.

The Royal Commission on the Sanitary State of the Army

The first British wide-ranging enquiry into health conditions within the indigenous population of India was the 1859 Royal Commission on the Sanitary State of the Army in India which closed with a report in 1863. While continuing to focus on the health of European soldiers in India, the commission's report included the Indian population in its wider investigation into the manifestation and causation of disease. The commission's findings confirmed British concerns for the health of Europeans in India, especially those in the army. The commission found that European army officers died at the rate of 31 per 1000 per annum, soldiers died at the rate of 69 per 1000 and European civilians died at the rate of 20 per 1000 per annum. The mean lifetime of Europeans resident in India was found to be shortened by almost 22 years.

The commission was significant not only as an indication of increased and wider-ranging British government interest in disease in India, but also for its role in formalizing a structured and comprehensive bureaucratic approach to government investigation of disease and the collection of statistical information on matters of public health. The development of systematic statistical investigation of disease in India during the 1860s was directly linked to the new attention to sanitation.[19] The Government of India strongly criticized the Royal Commission's report for overstating the death-rate of the British army in India by including war casualties, and for lack of recognition of the Indian government's efforts to improve the health and conditions of the European soldier in India prior to the commission's enquiries. Even so, the Royal Commission became the basis for greater British government attention to sanitary conditions in India. Some members of the British medical profession were sceptical about the degree to which sanitary measures could be regarded as scientific at all, their application being considered more 'a matter of method and handicraft, not of science'. However, this view did not prevent the *Lancet*, the pre-eminent British medical weekly of the nineteenth century, hailing the dawn of sanitary science in India as an officially endorsed and effective means of combatting disease and death.[20]

The principal medical initiatives in the wake of the Royal Commission were measures introduced in the 1860s and 70s to address the main sources of ill health and mortality in the army, namely cholera, venereal disease and excessive drinking. In the area of leprosy enquiry and treatment, rather than any specific medical initiative, the appointment of sanitary commissioners in Bengal, Bombay and the Madras presidencies in 1864 with partly consultative and partly administrative duties encompassing the civil, military, medical and engineering aspects of sanitation, was to have the most impact. Further, the Royal Commission indicated a change in epistemology. The kinds of questions raised by the commission – specifically, whether health was affected by locality, climate, facilities in barracks, drainage, water supply, diet, drink, dress, duties or habits, stripped of their military context, were applied not only in general sanitation enquiries but in the later enquiries into leprosy conducted by the Indian government's sanitary department.[21]

Medical journalism in Madras

Although from the early nineteenth century the scientific principles of observation, experiment and inference had gradually displaced the more speculative, humorally based forms of diagnosis and description of disease in European medicine, the culture of scientific enquiry into disease did not consolidate in Europe or India until the mid-nineteenth century. By the late 1860s, the *Lancet* acclaimed the rapid progress of science in India, despite the persistence of such obstacles to scientific medical research as the paucity and isolation of European medical officers, the punishing climate, the 'peculiar prejudices and feelings' of the Indian peoples which 'interfere very materially with the accuracy of clinical and physiological observation' and the lack of medical societies and medical journalism.[22] The picture painted by the *Lancet* was not, however, entirely accurate. The 'prejudices and feelings' of the Indian patients, though often a significant factor in resistance to British medical treatment of leprosy, posed less of an impediment to leprosy research than the 'prejudices and feelings' of the medical officers themselves. Further, by the 1860s, medical societies and journalism were providing an important forum, however modest, for the exchange of scientific medical ideas in India.

Madras had followed Calcutta and Bombay both in the formation of medical societies, and in providing a printed forum for medical discussion. The first professional medical periodical in India was the *Transactions of the Medical and Physical Society of Calcutta* issued in 1825, which ran under various titles until 1845, and was followed in 1834 by the first Indian medical publication not supported by a society, the *Indian Journal of Medical Science*, which continued until 1843. In 1838, their founding year, The Medical and Physical Society of Bombay began publishing their *Transactions* which continued into the twentieth century. Madras followed closely, with its first true medical journal, the *Madras Quarterly Medical Journal*, running from 1839 until at least 1843, and the *Madras Journal of Medical Science*, the first medical journal in India not edited entirely by members of the IMS, published from 1851–54. In July 1860, the *Madras Quarterly Journal of Medical Science*, edited by W.R.

Cornish and G.B. Montgomery, began publication, becoming a monthly in 1869 and continuing publication until 1873.[23]

The emergence of a specialist medical journalism in India in the 1830s, contributing to the increasing professionalization of medicine by the 1860s, was roughly contemporaneous with efforts to raise the prestige of the medical fraternity in nineteenth-century Britain. In Britain in the 1830s there was a similar development of medical journalism. In 1858, the Medical Registration Act marked a further stage in the medical profession's efforts to attract the authority and status which continued to prove elusive. In mid-Victorian Britain, the low socio-economic status of many medical practitioners and the upper-class association of medicine with 'quackery' persisted. The growth of British medical professionalism was fuelled not so much by regulation and training, as Bala suggests,[24] but by increasing commitment to the new scientific medical epistemology.

While preoccupation with status often prompted competition rather than co-operation between provincial medical services, in the 1860s it also served to stimulate medical journalism in Madras. The publication of medical journals reflected an effort within the Madras Medical Service to carve a status for itself at least equivalent to that held by the older Bengal and Bombay medical services. As Samuel Rogers, assistant surgeon, noted in the preface to the first volume of the *Madras Quarterly Medical Journal*, the object of the publication was 'to contribute something to the general stock of professional information and at the same time to raise the character and respectability of the body to which we belong'. He hoped that the *Journal*'s publication of reports, monographs, cases and other materials would be regarded as evidence that, though later in origin than the Bombay and Calcutta medical services, the Madras service would follow their 'meritorious example' and 'that our contributions, like those of the Sister Presidencies, may lead to the advancement of the medical art'. Further, by publication of official statistical and medical reports compiled for the army, the editor hoped to cultivate an intellectual and scholarly approach to medical investigation among the medical officers in Madras in 'meritorious emulation' of the medical profession in Europe.[25]

By contrast with the legal service, the display of merit through publication gave little assistance to promotion at the senior levels of the medical service. Even so, medical societies and, more importantly, specialist medical journals, did contribute to the professionalization of medicine and gave new opportunities for the discussion and publication of the results of individual medical investigation. For members of the Subordinate Medical Service, publication was seen as an opportunity to raise their status and win the respect of their superiors in the Indian Medical Service. The *Madras Journal of Medical Science*, published from 1851 to 1854, was entirely conducted by members of the Medical Subordinate Department of the Madras presidency and drew its primary contributions from that department. The principal object of the journal was to enable 'intelligent studious Medical Subordinates' of the Madras presidency to display their 'reading, experience, and observation, in the great Field of Disease', to show their capacity for professional development and to earn for their class greater respect from their superiors by indicating through publication, 'their zeal, intelligence and perseverance'.[26]

The publication of the *Madras Quarterly Journal of Medical Science* in 1860, in a climate of sustained medical journalism, was indicative of the greater confidence of the Madras medical profession by the 1860s. While the *Madras Quarterly Medical Journal* began publication in 1839 with a subscription list consisting almost entirely of British members of the Madras Medical Service, the *Madras Quarterly Journal of Medical Science*, in addition to a wide-ranging subscription list, boasted a list of publications for exchange with the journal which included not only the principal medical journals of the other Indian presidencies but those of Britain and America.[27] The Madras Medical Service was sure enough of its professional credentials to face the world.

From the mid-1860s a further boost to the professional status of the Madras Medical Service was to be provided by the increased number of recruits who held medical degrees and were fellows of the Royal Colleges. From 1865 the superiority of the Bengal Medical Service over the Bombay and Madras services in terms of its members' qualifications was significantly reduced, even though, since Bengal offered unparalleled opportunities for professional advance and lucrative private practice, the Bengal service never lost its attractiveness to the best IMS recruits.[28]

By the 1860s the British government had begun to consider the health not only of the British, but also of the Indian population to be part of their responsibility. Coherent government enquiry into disease and the notion of sanitary science were established as valid forms of medical involvement with the Indian people. Further, a scientific medical culture was rapidly developing. Medical societies and medical journalism provided a forum for discussion and dissemination of the scientific principles of medical research.

The development of a scientific approach to leprosy investigation from the 1860s was both facilitated by and contributed to the growing culture of scientific medical enquiry in India. The *Madras Quarterly Journal of Medical Science* provided considerable stimulus to medical attention to leprosy in Madras. Day's discussion of the symptoms of leprosy and the nature of the disease was published in 1860, and van-Someren's description of leprosy at the Madras Leper Hospital, under his charge, followed in 1861. Both papers were based on the detailed observations of the physical symptoms of leprosy as evident in those under the author's care. A brief article: 'Notes of a case in which a Leper was attacked with Small-pox' by J. Shortt, MD, then zillah surgeon at Chingleput, but made superintendent-general of vaccination in 1865, appeared in the journal in 1862 as a consequence of these articles.[29]

Post-mortem leprosy research

The first Indian anatomical exploration of leprosy by means of autopsy and microscopic examination, entitled 'On the Symptoms and Morbid Anatomy of Leprosy, with remarks', was published through the Medical and Physical Society of Bombay in 1865 by H.V. Carter of the Bombay Medical Service. Carter's work was stimulated by Danielssen and Boeck's pioneering research into the anatomical characteristics of leprosy, but went further, employing microscopy to examine the minute structural changes in the nerves caused by the disease.[30] When Carter's 'Morbid Anatomy' went to press there was still little information on leprosy published by the medical fraternity in India and a similar paucity of texts world wide. However, medical societies and journalism were making medical research readily available. Carter had access to Danielssen and Boeck's work through a detailed review of *Om Spedálskhed* in the *British and Foreign Medico-Chirurgical Review*.[31] Indicating the calibre of the new research from Madras, Carter cited Day and van-Someren's papers as among the most significant recent contributions to the understanding of leprosy.

Carter's work, like that of the Madras surgeons, was based on the direct observation of leprosy patients. In Bombay, Carter observed more than a hundred patients each year, suffering from various stages of leprosy, who were admitted to the Native Hospital and Male Dispensary at the Jamsetjee Jejeebhoy Hospital or cared for by the

Dhurumsala, an asylum for the homeless, under the auspices of the local District Benevolent Society. His post-mortem research was carried out on those leprosy sufferers who died at the hospital. Carter's use of hospital bodies for post-mortems reflects the low status of leprosy sufferers at the hospital. Most were likely to be paupers without friends or family to ensure their proper disposal according to custom. It is possible though, that some who died with leprosy could not be cremated according to Hindu rites. The strength of Indian opposition to providing corpses for dissection meant that prisons were one of the few British sources of bodies for medical research.[32] The leprosy sufferer who died in hospital without home or friends was, in death, as much at the disposal of the British medical service as the prisoner.

Carter's enthusiasm and success in performing post-mortems was unusual in India and reflected his particular interest in morbid anatomy. While demonstrator of anatomy at St George's Hospital, London, he had collaborated with Henry Gray on the dissections and drawings for the woodcuts which illustrated the famous *Gray's Anatomy*. Although some medical officers did perform autopsies for research using bodies from prisons or hospitals, there were particularly strong impediments to post-mortem examination of leprosy sufferers. When 'Morbid Anatomy' was published, Morehead's general work *Clinical Researches on Disease in India* was the only text offering any post mortem examinations. Shaw observed in 1864 that the paucity of 'satisfactory *post-mortems* of lepers' performed in the Madras presidency was largely due to the 'loathsomeness of the disease, the heat of this climate, [and] the prejudice of the Natives'. The increased medical interest in leprosy from the 1860s did, however, draw attention to the technique of post mortem examination of leprosy sufferers.[33] Asylum and hospital populations of leprosy sufferers were beginning to be seen not only as a context for charitable intervention by the medical service but, like jails, as a medical and scientific resource.

The publication of their findings on leprosy by van-Someren and Day in Madras, and Carter in Bombay, introduced a new professionalism to the study of leprosy in India and a more empirical approach to the investigation of the disease. Carter advocated strict adherence to 'direct and methodical observation' of physical signs as the new means of understanding the nature of leprosy and expressed his hope that the 'modern' medical writers would make the next step in the knowledge of leprosy 'aided by means of investigation till now unexplored'.[34] The publication of Day and van-Someren's articles in a journal which anticipated a distinguished Indian and international medical readership, indicated a shift from the earlier local charitable medical efforts by IMS officers in Madras to ease the leprosy sufferers' situation, to a more professional medical and scientific response developed in the context of the Indian and international medical profession. For van-Someren, as for Carter, the main stimulus for publication of systematic and methodical research into leprosy was not local but European.

European influence and scientific medicine

Like Carter's article, which had been inspired by work on leprosy in Norway, van-Someren's 1861 article on leprosy in the Madras Hospital was prompted by European attention to leprosy. Van-Someren's article was initiated in response to an enquiry into the aetiology, distribution, cure and control of leprosy published by Professor Rudolf Virchow of the University of Berlin. In 1859, Virchow had been invited by the Scandinavian government to study the etiology of leprosy in Norway. Although not entirely successful in terms of research, the journey proved of great significance in disseminating knowledge of the disease. Virchow appealed to physicians, historians and travellers of all countries for information to assist him in composing a history of leprosy, by then clearly identified as *Lepra Arabum, Elephantiasis Grœcorum*. The types of questions asked by Virchow indicate the dearth of information on leprosy in India at the time and the type of thinking on leprosy developing in Europe.

Virchow's questions comprised two parts: Part A ascertaining the number of leprosy hospitals available, and Part B seeking information on the disease itself and placing leprosy in the context of measures of disease control. Part B requested information on the distribution and prevalence of leprosy, the forms of the disease observed, whether the disease was endemic or sporadic, its possible causes, including inheritance, contagion, climate (particularly the influence of humidity), the eating of certain foods (especially fat or fish), and whether there was any known treatment for the disease. Further, Virchow enquired whether there were 'any peculiar laws affecting lepers' such as solitary confinement or prohibition of marriage between those with leprosy.[35] Virchow's questions expressed the ways in which thinking about leprosy was developing internationally and the kinds of issues concerning means of control which were to dominate the late-nineteenth-century debates on leprosy confinement.

Virchow's enquiry had already attracted favourable attention in Britain and internationally, being publicized in the *Lancet* of July 1860 and brought to the notice of the International Statistical Congress in the same year. The first responses to Virchow's appeal, which described the conditions of leprosy in Germany, were published in the 18th volume of Virchow's *Archives*. In Madras, a reiteration of the appeal dated 19 April 1860 was published in the *Madras Quarterly Journal of Medical Science* of 1861. In response, both the journal and the Madras government co-operated in drawing the Indian medical profession's understanding of leprosy into the international context. Information could be forwarded by the journal or be sent in an official form directly to the principal inspector-general of the Madras medical department.[36] While the stimulus for obtaining medical knowledge on leprosy in India came from the emerging understanding of leprosy in Europe, the response of the medical fraternity and the Madras medical department indicated that, by the 1860s, there were both the structures for communication and publication available, and the willingness to devise an informed medical response.

The successful investigation of leprosy by Carter and van-Someren demonstrated the level of scientific investigation of leprosy possible by those medical officers who were willing to approach leprosy sufferers. The developing professionalization of medicine in the presidencies gave an incentive for the publication of their findings and facilitated the development of a medical literature on leprosy based on empirical method. Even so, research into leprosy was pursued principally by those medical officers whose particular interest in the leprosy sufferer's welfare was sufficient to overcome revulsion of the disease.

The Royal College of Physicians

The first major medical investigation, presidency by presidency, of leprosy in India sponsored by the Home government was initiated by Jas Walker, commander-in-chief of the Windward Islands, Barbados. Walker recommended to the Colonial Office that reports be made on leprosy in the West Indian colonies and collated by some professional medical body in England. He was convinced that such an enquiry was needed to provide information for the study and guidance of the colonies in dealing with leprosy since in the Windward Islands there was little knowledge of the disease to inform a response to the apparent increase in its prevalence. There, as in India, research into leprosy was considerably impeded by repulsion.[37]

The colonial secretary, the Duke of Newcastle, secured the cooperation of the Royal College of Physicians (RCP) of London, who appointed a leprosy committee consisting of college fellows to frame 'interrogatories' for circulation and to collate, digest and report on any information collected. The leprosy committee consisted of Dr Budd (Senior Censor), Dr Owen Rees, Dr A. Farre, Dr Gull, Dr Milroy and Dr Greenhow. Conclusions on the evidence provided for the RCP *Report on Leprosy*, were made by the leprosy committee together with Thomas Watson, the president of the college, and Henry A. Pitman, the college registrar.[38] The procedure of sending out interrogatories was a relatively new method of gaining information on the health and welfare of a population, part of the 'investigative tradition' developing in Britain from the 1830s. Starting in the midnineteenth century, a clear trend developed in Britain away from individually led health-reform action towards the formation of societies which could gather information and act as pressure groups.[39] Not all in the British medical profession were convinced of the value of statistics, some objecting that they were frequently inaccurate and at best no more than glorified 'returns'.[40] The cry 'lies, damned lies and statistics' could already be heard in England. Even so, as with the Royal Commission on the Sanitary State of the Army, despite criticism, the notion of statistical enquiry prevailed and contributed to the development of a culture of scientific medical enquiry in India.

It is suggestive of the lack of British government interest in leprosy that the Duke of Newcastle was unaware of the existence of the disease in any colonies other than the Windward Islands, and that he relied on the Royal College of Physicians to possess such information and to determine which colonies should receive the interrogatories. On the college's recommendation, the Colonial Office extended the investigation to include the entire colonial possessions of the crown,[41] and the Royal College of Physicians' enquiry became the first official British governmental and professional investigation of leprosy in India. The 18 interrogatories drawn up by the college were similar to Virchow's questions, indicating the consistency of the college's thinking on leprosy and control of the disease with emerging international medical concerns. British ignorance of the health and well-being of their colonial subjects was clear from interrogatories II–V which requested basic demographic information regarding the gender, race and age at maturity of those with the disease. While at the local level Europe was providing the stimulus for medical investigation of leprosy in India, the British medical profession, under the auspices of the Colonial Office, was setting the agenda for government enquiry into leprosy. By the mid-nineteenth century, in the area of leprosy investigation, colonial medicine at the periphery was being more closely driven by metropolitan medicine at the centre than it had been in the first half of the century.

The Royal College of Physicians Leprosy Committee's *Report on Leprosy*, published in 1867, had a dual aim. The first was to gather information for the Colonial Office on the contagiousness of leprosy and thus provide a guide to the appropriateness of legislative action to confine leprosy sufferers. The second was to furnish the medical profession with 'a better insight into the causes, treatment and prevention of an almost unconquerable malady'.[42] Published in the same year that the *Lancet* claimed medical science to be 'making rapid strides in India', the committee's report made a significant contribution to the collection and distribution of information on leprosy, and ushered in a new period of methodical and 'scientific' investigation of the disease by British medical and government authorities, who were driven partly by a sense of imperial responsibility. The report noted that, in India, prior to the college's enquiries, leprosy had 'excited but little attention either in a scientific or social point of view' and expressed their hope 'that the present enquiry may lead to a more thorough and systematic examination of a malady which affects so deeply the material well-being and interests of millions of our fellow-creatures, subjects of the British crown'.[43]

The degree of interest in leprosy generated by the Royal College of Physicians' enquiry was evident in its considerable print run of 3000 copies. In India, the college's enquiries did provide a strong stimulus to investigation into leprosy. Of over 250 replies to the college's interrogatories, not including those of crown consuls and governors of British colonies, more than half came from India.[44] The British medical profession had high expectations of the report's contribution to the development of scientific medicine. The *Lancet* emphasized the report's value as 'inductive research' and anticipated that the material collected would not only 'throw much light on a very obscure disease', but prove 'an important addition to medical science'.[45]

New medical knowledge

The Royal Commission on the Sanitary State of the Army and the Royal College of Physicians' enquiry marked a departure from the medical topography reports compiled by medical officers for submission to the presidency governments in the first half of the nineteenth century. Quite distinct from the gentlemen-scholars' collaborations on indigenous remedies, the medical topographical reports were broad surveys including anecdotal information on environment and living conditions of both the British and indigenous peoples in India, and returns quantifying diseases and treatments in different localities. The reports were proposed in 1835 by Ranald Martin in response to the debilitating effects of fever and cholera on the British troops in India. They were intended to assist British efforts to understand the physical environment of India and thus make the country less medically dangerous, more lucrative for the European occupants, and, particularly, safer for the army.[46]

From the 1860s, broad medical topographical surveys were supplanted by more specialized forms of medical enquiry. This specialization not only reflected the tendency for bureaucratic proliferation as crown rule expanded and consolidated, but also the development in Britain, and adoption in India, of statistical enquiry and sanitary science as a means of understanding and controlling disease. In India, in addition to reports commissioned to address particular diseases, such as the statistical surveys of the 1860s and 1870s stimulated by sanitary and cholera enquiries, the routine recording of public health data was institutionalized and, from 1871, incorporated into the wider social and economic investigative structure of the all-India census. It has been argued by Mackenzie and others that the expansion of statistical enquiry in Britain was, in part, an effort to know and reform the pauper population.[47] In India, from the early nineteenth century, British concern to know and thus adapt to and manage the Indian environment prompted not only the medical topographical reports, but also geological, botanical, zoological and meteorological surveys. Recent writers on medicine in India have tended not to argue for medicine as a means of 'surveillance' and 'social control' of the population, emphasizing rather the limitations of medicine and public health measures as a form of surveillance. Even so, in the early nineteenth century vaccination against smallpox did provide a means, albeit imperfect, of identifying and locating the Indian population, who were largely unknown to the British colonizers.[48] In the context of the Royal College of Physicians' enquiry, the need to 'know' the population in order to respond to disease was evident, as was the potential for disease-control measures to act as a means of controlling certain sections of the population. The first aim of the college's leprosy investigation was to determine, on the basis of the contagiousness of leprosy, whether or not to legislate for confinement of leprosy sufferers.

Despite the increase in British and Indian government interest in leprosy, the acquisition of knowledge in a form appropriate to the requirements of 'direct and methodical' observation, was still very much dependent on local IMS interest in leprosy and willingness to investigate and report on findings. Even though endorsed and facilitated by both British and Indian governments, the college's enquiry was conducted at the local level by medical officers already involved in leprosy work. There were no investigators sponsored exclusively by the British government, as were later employed for the 1890 Leprosy Commission, and the medical officers received no additional payment for their research and formulation of replies to the interrogatories. The quality and significance of the information gathered by IMS officers was highly regarded by the college's Leprosy Committee, and on general questions of the character of the disease the Indian reports on leprosy, especially Carter's research, were given prominence in the Report's conclusions.[49] Medical investigation of leprosy at the colonial periphery had, by the late 1860s, become sufficiently compatible with medicine as practised at the centre of empire, to be incorporated in its official medical response to leprosy.

Colonial medical science in India was, by that time, not simply perceived by the medical profession in Britain as a form of '"derivative" science, done by lesser minds working on problems set by *savants* in Europe', as MacLeod's definition of 'colonial science' would suggest.[50] Consistent with Basalla's model of the expansion of European science, leprosy investigation in India during the phase of 'colonial science' was still largely dependent on European initiative. However, at the same time, a more independent Indian metropolitan science, following the European method, was becoming well established, and the work of some practitioners of medical science in India had sufficient authority in Europe to be placed alongside Danielssen and Boeck's research.

'The interests of science and humanity'

Home government responsibility for leprosy investigation

The Royal College of Physician's Report on Leprosy urged the British government's continued commitment to the investigation of leprosy in the colonies and signalled college solidarity with the government in an endeavour considered to be both scientific and humanitarian. The Leprosy Committee commented that:

> The great extent to which leprosy prevails in many distant dependencies of the British empire, and the inevitable destitution and mendicancy that attend its existence among a population, render its thorough investigation a matter of special duty on the part of the Government of this country.

The college's assertion that the investigation of leprosy by other colonial authorities, notably France, Spain, Portugal and Holland, would be beneficial to both 'the interests of science and humanity'[51] suggests that Britain styled herself world leader in both moral and scientific terms and, by implication, affirmed the college's medical superiority in the field. Actually, in comparison with Norway, Britain was backward in attending to the care of leprosy sufferers and the control of their disease. From 1832 to 1856, Norway had conducted four national leprosy censuses, instituted a permanent national registry of leprosy sufferers, funded clinical study of the disease and in-patient treatment, implemented a national scheme for the control of the disease and developed a valuable theory of leprosy pathogenesis. However, in terms of government-endorsed colonial exploration of leprosy, the British were ahead. The Dutch were the first to follow the British, publishing a study of leprosy in Dutch Guiana, entitled *De la Contagion, seule Cause de la Propogation de la Lepre*, in 1869.[52]

The Royal College of Physicians' report clearly linked medical science with humanitarian reform and endorsed government-supported research into leprosy. The impulse towards charitable medical assistance of the leprosy sufferer, evident in Madras in the first half of the nineteenth century, was to continue, not simply as a local phenomenon but rather, with the support of the government and the medical profession in Britain, becoming bureaucratized and integrated with the pursuit of scientific interests. The more methodical and scientific approach to leprosy investigation did not, however, result in disregard, at the local level, for the leprosy sufferer's experience of the disease. Day's description of the effects of anaesthetic leprosy, expressed well the local medical feeling:

> Probably a more distressing sight does not exist than a poor leper crawling about on only the stumps of his hands and feet, destitute of eyesight, and covered with sores. When death from exhaustion, dysentery, diarrhoea, bronchitis, or dropsy terminates the earthly career of such unfortunate remnants of humanity, it must be more a subject for congratulation than either pity or regret.[53]

The Royal College of Physicians' *Report on Leprosy* and the professionalization of medicine in England

Somewhat more removed than Day from the predicament of the leprosy sufferer, the Royal College of Physicians' concern for the well-being of empire, and its identification with the government in commitment to the empire's care, was not entirely founded on altruistic or scientific concerns. Rather, it was substantially motivated by the college's interest in improving its professional status by association with government endeavour. From June 1862 to February 1867 the *Lancet* paid close attention to the instigation and progress of the Leprosy Committee's enquiries and, in 1866, strongly endorsed the British government's action in appealing to an accredited professional medical body for medical knowledge and expertise rather than seeking advice from private individuals.[54] This approval reflected the same concerns which had precipitated the Medical Act of 1858: principally, a need to stipulate what medical qualification entailed and to confer on a rather disparate collection of medical practitioners a coherent status and identity as members of a profession. The Royal College of Physicians was the most prestigious medical corporation in England and the oldest, its charter dating from 1518. Its members, the practitioners of 'physic' (internal medicine), were distinguished from members of the Royal College of Surgeons and the Society of the Apothecaries by their university education. The Royal College of Physicians monopolized the practice of 'physic' in London and its environs, and was generally responsible for physicians in all of England. It was concerned to maintain its superior status in the atmosphere of increasing unity and professional medical reform following the Medical Act of 1858,[55] and the Leprosy Committee's report made an important contribution to the college's medical authority. In January 1867, the *Lancet* asserted that the Royal College of Physicians' Report 'must be looked upon as *the authority* on leprosy' and anticipated with some satisfaction that after the Leprosy Committee's success, consultation between the college and the British government would become more frequent. The following month the report was held up as 'an excellent sample of the kind of information which the government might get by listening to scientific men on scientific questions'.[56] Almost a decade after the Medical Act, the British medical profession expected the Leprosy Committee's report to be as beneficial to themselves as to the leprosy sufferers in British colonies.

The influence of college members on the development of a scientific approach to leprosy in India was not limited to the Royal College of Physicians' 1867 report. Tilbury Fox, an eminent member of the college, together with T. Farquhar, devised 'A Scheme for Obtaining a Better Knowledge of the Endemic Skin Diseases of India'. The 'Scheme' was recommended to the British government by the Army Sanitary Commission and in 1872, on the order of the Duke of Argyll, secretary of state for India, distributed to medical officers in India with an interest in the subject, along with a request for information on the cutaneous and endemic diseases listed, which included leprosy. Subsequently, at the request of the Colonial Office, the scheme was circulated to medical officers in other foreign and colonial stations and the results of the enquiry published by the India Office in 1876 as *On Certain Endemic Skin and Other Diseases of India and Hot Climates Generally*.

Fox and Farquhar's intention in distributing the 'Scheme' was to obtain a better knowledge of the more significant endemic diseases of India which principally attack the skin, and to circulate the information obtained. The combination of a newly scientific approach to leprosy investigation with the charitable attitude to leprosy care, apparent in India from the 1860s, continued with Fox and Farquhar's project. The authors anticipated that their work would not only 'come to be regarded as a really solid and valuable contribution to medical science' but would advance understanding 'of a class of very distressing and very obstinate diseases'.[57]

In addition to its general contribution to the development of medical science, Fox and Farquhar's 'Scheme' and subsequent report were intended to contribute specifically to the development of medical science in India. The 'Scheme' was more clearly focused than the Royal College of Physicians' enquiry, which had aimed to offer the medical profession 'a better insight' into leprosy. The 'Scheme' was intended to bring an agreement between the disparate opinions of the medical profession in India and England regarding the 'nomenclature, typical character, the varieties, and the probable or demonstrated causes of the diseases in question'. Further, it was hoped that the 'Scheme' would assist in standardizing and improving the training in understanding and treating common tropical diseases available for medical officers intending to work in India, there being little opportunity to gain clinical experience of these diseases in England. The enquiry was expected to be of reciprocal benefit, also enabling practitioners in England and on the continent to diagnose more accurately and treat more effectively the many cases of skin disease imported from India.

The details of the 'Scheme' were arranged in response to some of the particular obstructions to research experienced by Indian medical officers in the field. As with the college's enquiry into leprosy, the facts were to be gathered by 'the scattered and able medical officers of India' and their opinion sought on their findings. Since it was still often impossible for medical officers in India to gain access to medical libraries and periodical publications recording the most recent European research, the latest information from leading European dermatologists on the several diseases listed was included in the 'Scheme'. The European information on leprosy was drawn principally from the Royal College of Physicians' report, indicating its continued status in Britain as the authority on leprosy in India. Recognizing the difficulties encountered by Indian medical officers in carrying with them 'the necessary apparatus for minute and experimental enquiry and the like', Fox and Farquhar defined the points of doubt requiring elucidation in order to economize on the time and labour required to formulate a response to the 'Scheme'. They also proposed specific lines of further enquiry into the aetiology of the diseases in question.[58]

Fox and Farquhar's work was not only intended to facilitate research in the short term but provided specific and deliberate encouragement to the continuing development of a professional and scientific approach to the medical investigation of cutaneous disease in India. The authors were confident that their report's provision of accurate, information on current research would prove an essential service in the development of medical knowledge in Indian districts which 'afford few, if any, facilities for references to standard works or recent periodicals'. In 1873, the inspector-general of hospitals for the Indian Medical Department recognized that further information 'of value' on leprosy and other cutaneous diseases could not 'be expected to accrue immediately or even shortly' as a consequence of Fox and Farquhar's work. Even so, he highly recommended their report on endemic skin and other diseases to the Government of India and confirmed the value of their findings in providing an information base for further research.[59] Thus, by the 1870s an expectation had developed at the highest level of the Indian medical bureaucracy, that with cautious and patient application there would be a further advance in the acquisition of empirically based information on cutaneous disease in India, including leprosy.

Fox and Farquhar's 'Scheme' illustrates well the continued dependence of colonial science, as understood by Basalla, on European science. Further, it highlights the concern of some practitioners of metropolitan science at the centre to advance the establishment and quality of such science at the periphery, principally by providing information. By the 1870s, the Government of India, though offering no concrete assistance in terms of funding and scientific apparatus, was willing at least to recognize the value of scientific research into leprosy and other cutaneous diseases by facilitating the distribution of Fox and Farquhar's 'Scheme'.

E.G. Balfour, surgeon-general for the Madras presidency, commented that several years would need to elapse before the principal questions in the 'Scheme' could be determined by medical officers who were newcomers to investigating cutaneous pathology. Even so, Balfour anticipated that since special attention had now been directed towards leprosy, further information from the medical officers could be expected. The lack of new research on leprosy in the Madras presidency since 1860 was evident in the replies to the ' Scheme' published in 1875 as *Reports Forwarded by Surgeon-General E.G. Balfour on Certain Forms of Skin Diseases Observed in the Madras Presidency.* Apart from specific responses by the four medical officers reporting on endemic leprosy in their districts,[60] Balfour's reply consisted principally of the 1871 Madras census findings on leprosy, Shaw's 1864 report for the Royal College of Physicians, and reports on the treatment of leprosy with the cashew nut, gurjon and chaulmugra oils, which had recently come to the attention of the British and Indian governments. The inclusion of information on leprosy in the census marked the increasing attention paid to leprosy by the Government of India.

Balfour's report was perhaps most important for what it demonstrated of the status given to indigenous information on leprosy in the European 'scientific' medicine of the 1870s. Fox and Farquhar's 'Scheme' was the first Home-government-sanctioned investigation of leprosy in India to enquire specifically about indigenous understandings of the disease. The reports collected by Balfour indicated that although British medical officers were willing to employ indigenous remedies for leprosy, they were sceptical of the validity of indigenous conceptualizations of the disease. Van-Someren commented that the descriptions of leprosy in Tamil medical texts were 'quaint' and the classification of leprosy into 18 forms was 'fanciful and artificial', with some conditions described as leprosy being actually nothing of the kind.[61] Further, although requested, the information on indigenous understandings of leprosy was excluded from Fox and Farquhar's final report, 'On Certain Endemic Skin and Other Diseases'. From the position of 'metropolitan' medical science, as practised in both Europe and India, indigenous perceptions of leprosy were not only inconsistent with European ideas, but inferior and not warranting incorporation into the new scientific understanding of leprosy developing on the basis of Danielssen and Boeck's research. This marked a significant shift in attitude from that of the early nineteenth century when H.H. Wilson, in an article published in 1823 in the *Transactions of the Medical and Physical Society of Calcutta*, suggested that indigenous understanding of leprosy may offer guidance and lead to 'improved knowledge and classification, if not to a more successful treatment of the disease'.[62]

By the 1890s a scientific medical culture was well developed in both Britain and India, contributing to a loss of faith in both indigenous medical treatments and the value of indigenous conceptualizations of disease. Leprosy research, despite its impediments, moved from being almost exclusively a local matter into the international arena, contributing as much to the professional needs of medicine in Britain as to the understanding of the disease. A colonial science was emerging which, though closely connected to Britain, had its own dynamic.

6
The Politics of Leprosy Control

Introduction

From the early 1870s, both leprosy treatment and scientific investigation of the disease did not remain entirely altruistic but were driven increasingly by political concerns. During the 1860s the investigation of leprosy was used by the Royal College of Physicians to bolster its emerging professional status. However, the testing of Beauperthuy's cashew nut oil treatment in the early 1870s made clear the limits of the Royal College of Physicians' authority. With the subsequent development of the gurjon oil treatment by Dougall in India, the Government of India and the sanitary department took the lead, displacing the British government and Royal College of Physicians as directors of leprosy investigation in India. The attention paid by both the British and Indian governments to each cure marked a considerable departure from the situation in the early nineteenth century when treatment initiatives such as Dalton's came only to the attention of the presidency government. Beauperthuy and Dougall's treatments, each of which was hoped to be a specific cure for leprosy, were trialled in the Madras presidency as part of a wider colonial program of treatment trials, indicating not only the degree of British and Indian government interest but also the extent to which intervention in and regulation of leprosy investigation had developed. Increased government regulation of leprosy treatment did not, however, spell the end of local treatment control. Medical officers in Madras exercised discretion in implementing government trials, and indigenous remedies remained an important element in local British medical care for leprosy sufferers.

From the mid-1870s the international leprosy debate shifted as a result of conclusive evidence from Norway that leprosy was contagious and consequent Norwegian legislation for the isolation and confinement of leprosy sufferers. Under Norwegian influence the focus of leprosy investigation in India moved away from cure of the disease and towards containment of the leprosy sufferer. Again, the issue was highly political. The new sanitary department in India took on the role of the Royal College of Physicians and used the investigation of leprosy as a means to assert its professional superiority over the Indian Medical Service. Critical to the initiation and development of legislation for leprosy control in India were the struggle between the sanitary department and the IMS for control of information about leprosy; the fear of contagion after the death of Fr Damien de Veuster of Molokai on 15 April 1889; and the consequent 1890–91 Leprosy Commission to India.

Professional medical authority and Beauperthuy's cure

The Royal College of Physicians continued to play a prominent role in setting the British and Indian government's agendas for leprosy investigation during the 1870s. However, during discussions from 1869 between the British government and the Royal College of Physicians to determine the best means of investigating Beauperthuy's alleged cure for leprosy, it became apparent that there was a difference of priority in their approaches to leprosy investigation. Ultimately, the Indian government's testing and promotion of Beauperthuy's mercury and cashew nut oil remedy was initiated without regard for the college's objections.

Dr L.D. Beauperthuy, who held the distinguished degree of Doctor of Medicine of the University of Paris, had been resident in Venezuela for 30 years during which he had been engaged in the investigation of leprosy and other diseases found in the tropics. In 1868, Beauperthuy's apparently successful treatment of leprosy came to the attention of the government of Trinidad and thence to the Duke of Buckingham and Chandos, principal secretary of state for the colonies, through A.H. Bakewell, president of the medical board of Trinidad and visiting physician to the leper asylum there. Reflecting the continued importance of scientific exchange in the development of leprosy research, Bakewell had read of Beauperthuy's alleged cure in public journals and the official gazette of Neuva Andalucia. Bakewell had a particular interest in the alleged cure, having made a series of pathological, chemical and microscopic investigations into leprosy and written part of a clinical work on the disease himself. In other colonies, as in India, the investigation of leprosy and search for cure had continued independently of government. Among at least some members of the colonial medical fraternity there was both a belief that a cure would be discovered and a sense of competition in being the discoverer.[1]

The Royal College of Physicians retained a significant role in directing the British government's response to leprosy investigation in the colonies. The Colonial Office endorsed the view expressed by the college in their *Report on Leprosy*, that the British government had a responsibility for the welfare of the leprosy sufferers in its empire. The colonial secretary was angered by Beauperthuy's withholding of full details of his alleged cure for leprosy, stating that every week of delay 'involves a responsibility on his part for an enormous amount of needless human suffering in every quarter of the world'.[2] Prior to the appearance of Beauperthuy's cure, the college had developed a policy of refusing to recognize any medicine of which the nature and composition were undisclosed,[3] no doubt as an attempt to secure the college's reputation further against any associations with quackery.

The discussions between the Royal College of Physicians and the Colonial Office over the procedure for investigating Beauperthuy's treatment revealed a developing tension between the British medical and government authorities over who had ultimate authority in leprosy investigation. The Royal College of Physicians expressed their willingness to advise the government on any medical matter and, in particular, to offer the medical adviser of the Privy Council every possible assistance. However, they objected to the British government's failure to promptly follow their recommendation that a college-approved, independent assessor of Beauperthuy's cure be appointed.[4] Although the details of the cure had been disclosed in 1869 through Dr Bakewell's efforts under government direction, the college refused to endorse it without an assessment of its claims being made by someone other than Bakewell and suggested that the government's neglect of the college's requirements was a neglect of imperial duty.[5]

The Royal College of Physicians' caution was justified on medical grounds. Bakewell himself, particularly after a second observation of Beauperthuy's cure, strenuously denied that he had claimed that Beauperthuy had discovered a specific cure. He argued, both in his 1869 report to the Trinidad government and in a letter to the *Lancet*, that he had used the term 'cure' not in a scientific sense, but in a popular way to describe the disappearance of external leprous symptoms in those patients treated according to Beauperthuy's regimen. Bakewell emphasized that there was no specific cure as such but that by applying the different components of Beauperthuy's treatment regimen, the progress of the disease appeared to be arrested, at least temporarily.[6]

Only after the Royal College of Physicians had generally approved Beauperthuy's treatment, as drawn up by Bakewell, did the Colonial Office succumb to college pressure and request that the college name an appropriate person to continue the enquiry into leprosy in association with Bakewell. It was a threat from the college to reject the next government-endorsed report on Beauperthuy's treatment,[7] rather than their earlier suggestion of imperial negligence, a strategy perhaps proving more an irritation than an incentive, which finally won the Colonial Office's compliance. The Royal College of Physicians' caution at embracing Beauperthuy's claims for his undisclosed cure offers insight into the absence of British government endorsement for investigation and trial of other cures for leprosy which had been developed in the 1860s. Bhau Dajie's refusal to disclose his complete cure publicly was the principal reason for the lack of any widespread testing of the cure with either British or Indian government endorsement.[8]

The British government had less difficulty in following the Royal College of Physicians' recommendations which removed the financial burden of leprosy investigation from their books, than in accepting the college's demands regarding the appointment of a medical associate for Bakewell. Kimberley, on the recommendation of the college, argued that since the colonies stood to benefit from the results of further trials of Beauperthuy's treatment, those in which leprosy was most prevalent should bear the expense of the enquiry in proportion to their revenue. Part of the cost entailed the remuneration of Gavin Milroy, the Fellow of the Royal College of Physicians finally appointed by the British government at the college's request and with full approval of the British medical profession, to witness the testing of Beauperthuy's treatment in the colonies and determine its value as a permanent cure for leprosy. The certainty of financial support from at least some of the colonies was of sufficient importance to warrant the British government postponing the enquiry until assurance was gained.[9] According to the Royal College of Physicians' 1867 report, India was not among the colonies in which leprosy was of most frequent occurrence. Thus, although

Beauperthuy's treatment was trialled in India, India was not included among the colonies selected to provide funding.

The government's handling of Bakewell's description of Beauperthuy's remedy confirmed that the government and the Royal College of Physicians had different priorities in the approach to leprosy investigation and treatment. The British medical profession's confident expectation that the college's report would virtually guarantee that, in future, the government would rely on members of accredited medical bodies for guidance in scientific and medical matters was by no means fulfilled. In 1870, the Colonial Office requested Dr Bakewell to draw up a statement of method to be followed in testing Beauperthuy's cure in the colonies. Although the scheme was to be submitted for approval to the Royal College of Physicians before implementation, the Colonial Office, displaying an unprecedented degree of intervention in the medical aspects of leprosy investigation, requested its formulation without consulting the Royal College of Physicians, and then distributed it for testing without amendment, despite the college's objection to some aspects of the treatment. The instructions were highly detailed and exacting, the hygienic aspect of the treatment including the provision for each patient of a separate room, a mosquito net for the bed, separate bedding, clothing and utensils for eating and drinking. The college made no objection to Bakewell's instructions 'on general medical grounds'. However, they noted, on the basis of Danielssen and Boeck's experience with mercury and the evidence gathered for the college's report on leprosy, that the internal use of mercury, which according to Beauperthuy's regimen was to be taken twice daily in addition to the external application of cashew nut oil, had 'a decidedly injurious effect' on the patient's health.[10]

While the Colonial Office was principally interested in Beauperthuy's treatment as a leprosy cure and was guided in part by a sense of responsibility for empire as well as by financial considerations, the Royal College of Physicians was intent on applying principles of empirical method to ascertain the scientific credibility of the cure before staking their reputation on its endorsement. In addition to a genuine concern for the pursuit of imperial duty, the college was intent both to keep within their purview the determination of the cure's efficacy and safety, a task which they regarded as of considerable importance,[11] and to retain their status as influential medical authorities in leprosy investigation established with the 1867 report. The *Lancet* considered that the Royal College of Physicians' status before the medical profession had been confirmed by the college's 'distinct and authoritative' evaluation of the value of Beauperthuy's alleged cure on the basis of a final two-year trial on Kaow Island. The authority and status of the medical profession in Britain as 'gentlemanly' and scientific was more secure.[12] Even so, the direction of investigation in the colonies of a possible leprosy cure, with its potential for attracting the kudos of success, provided a further opportunity to distinguish the Royal College of Physicians as the expert body of medical specialists on whom the highest levels of government could rely for guidance, even if the independence of the Colonial Office made the position hard won.

Already, by 1871, the *Lancet* doubted that a specific cure for leprosy could be found. One columnist remarked that Beauperthuy's cure, like carbolic acid, and the large doses of quinine and purgatives recommended by Tilbury Fox, would not themselves be sufficient to cure the disease. Rather, he argued that only within the wider curative context of good hygiene and good diet could such additional therapeutic measures hope to effect a cure. Even so, in the following year the Government of India requested that local governments trial and report on Beauperthuy's treatment. In Madras the treatment was trialled by van-Someren at the Madras Leper Hospital. Although the high degree of British government intervention and financial and administrative regulation of the trials was accepted by the Indian government, at the presidency level, medical officers did not comply with the treatment regimen where it was not in the best medical interests of their patients. Medical officers in the Madras presidency tended to concur with the Royal College of Physicians in their criticism of the remedy. E.G. Balfour, inspector-general of hospitals, supported van-Someren's refusal to employ either mercury or alkalies recommended by Beauperthuy, arguing that van-Someren's action would be supported 'by everyone at all acquainted with the pathology and treatment of leprosy' since either treatment would 'do much more harm than good'.[13]

While the increased attention to leprosy at British government level was related to a sense of responsibility for empire, tempered by financial considerations, at the local level the investigation of leprosy treatment, though stimulated by government directive, was grounded in a more immediate sense of responsibility for leprosy patients. Although the British, and hence colonial governments, could ignore the Royal College of Physicians' caution about the use of mercury in leprosy treatment, those committed and interested medical practitioners directly attending to leprosy sufferers could not. The Royal College of Physicians' medical priority in the investigation of leprosy treatments was shared at the local level of medical colonialism in India, even though initially it was overridden at the more distant British government level. Beauperthuy's cure was abandoned by the Government of India after treatment trials in Madras and other presidencies had found that its capacity to ameliorate leprous symptoms fell far short of a complete cure of the disease, and that the pain of the treatment far outweighed its benefit.[14] Local medical opinion still had considerable influence in the ultimate direction of medical treatment for leprosy sufferers in Madras – more influence, apparently, than the Royal College of Physicians.

The testing of Beauperthuy's cure in the colonies indicated not only a new degree of interest but also intervention and regulation of leprosy investigation by the British government, it being the first occasion when a leprosy cure received widespread testing under British government direction. As the struggle between the Royal College of Physicians and the Colonial Office demonstrated, the integration of medical factors into the higher levels of government policy was usually secondary to the economic and political requirements of government. The pressure for policy based on medical concerns usually came from the medical specialists and those involved in medical care at the local level. Greater government involvement in the medical investigation of leprosy did not necessarily result in the formulation of government response based on the medical aspects of the disease.

Indian government initiative in treatment trials

Dougall's gurjon oil treatment

The gurjon oil treatment developed by Dougall at Port Blair, which followed closely on the heels of Beauperthuy's remedy, was the first instance of a cure developed in Indian territory receiving a high degree of Indian government attention, together with endorsement in both Britain and Europe. The Government of India responded vigorously to Dougall's reports of success,[15] resolving in 1874 that the remedy be trialled throughout India. Introduced at the Madras Leper Hospital in February 1874, the fame of gurjon oil treatment spread rapidly. Like Dalton's cure in the early nineteenth century, gurjon oil attracted so many leprosy sufferers that extra shelter had to be provided at the hospital.[16]

From within government the remedy was given considerable support by the sanitation department which was to become a principal player in the direction of leprosy investigation in India from the mid-1870s, asserting a role in India similar to that of the Royal College of Physicians in Britain. J.M. Cuningham, sanitary commissioner with the Government of India, strongly supported Dougall, describing gurjon oil in 1875 as the 'most useful agent ever employed for the cure of this loathsome disease'. In the same year, gurjon oil was promoted not only through the colonies but also internationally, specifically to Norway which from the time of the publication of Danielssen and Boeck's work had been the principal source of information and guidance in the treatment of leprosy both in Europe and India. Two 500-lb-capacity barrels of gurjon oil and Dougall's reports on the treatment were forwarded to the Norwegian government in the expectation that they would be disposed to give gurjon oil a trial.[17] For the first time, India was leading Europe in responding to leprosy. The gurjon oil remedy was the first instance of an independent initiative by a member of the Indian Medical Service not only receiving a strong medical response within India but also Indian government support and promotion without instruction from Britain or Europe.

Discussion of gurjon oil as a leprosy treatment was not confined to government circles. Dougall's reports on the treatment of leprosy with gurjon oil were offered by the surgeon-general for publication in the *Indian Medical Gazette*, appearing from early 1874.[18] A lively public debate concerning the treatment's merit ensued through both the general public and medical press in India. The early medical response was generally optimistic, though cautious about claiming the oil as a specific cure for leprosy.[19] In Britain, the medical press was similarly optimistic. At St Bartholomew's Hospital, London, a person suffering leprosy was reported in the *Lancet* in June 1874 as having improved after treatment with Dougall's remedy though it was considered to be too early to determine its true effectiveness against the disease. By December 1874, however, the *Lancet* had become even more reticent in claiming gurjon oil as a new cure, describing it rather as the remedy most able to arrest and mitigate leprosy.[20]

In India, Surgeon G.C. Roy offered the most cogent criticism of the gurjon treatment, suggesting that even after two or three years continuous application, the gurjon oil produced no other change than 'a temporary improvement in the leprosy sufferer's constitution'.

Roy's claims attracted a heated defence of Dougall and the gurjon oil from Dougall's assistant apothecary at Port Blair, G.W. Phillips.[21] However, Roy's opinion proved the most correct and by 1876 it had become clear to Dougall that the gurjon oil remedy was not the specific cure he had hoped for, a view confirmed by the reports of mixed and often impermanent results of the treatment under trial. In 1877, the Government of India advised the local governments and administrations that 'the very hopeful prospect which the gurjon oil treatment held out has not been realised'.[22] In 1879 Cuningham, still sanitary commissioner with the Government of India, dealt the treatment a less diplomatic blow: 'The gurjun oil treatment is never heard of now. I am afraid it never merited the attention it attracted, but its author, Dr Dougall, is dead, and nothing more need be said about it'.[23] Even within the sanitary department, however, there was opposition to this view and in the medical profession, both in Britain and India, discussions of the oil did continue in the hope that it could offer at least some benefit to the leprosy sufferer.[24]

Local medical control

Johnston's carbolic fumigation

Despite the attention given to the cashew nut and gurjon oil treatments by the British government and Royal College of Physicians, investigation and trial of other British-developed leprosy treatments continued to be conducted at the local level in the 1870s, even to the neglect of treatment trials ordered by the Government of India. Deputy Surgeon-General W. Johnston introduced fumigation with carbolic acid vapour into Madras in the institutions under his jurisdiction, namely Palliport Lazaretto and Calicut Leper Hospital, beginning trials in 1872 and July 1873 respectively.[25]

Johnston's trials underlined both the degree of development in scientific culture and understanding of leprosy by the 1870s and the persistence of local medical control over leprosy treatment. Johnston's theory of leprosy causation and the physiological changes wrought by the disease was grounded in recent developments in the understanding of leprosy, principally based on Danielssen and Boeck's research, the observations contained in the Royal College of Physicians' *Report on Leprosy*, Milroy's *Report on Leprosy and Yaws in the West Indies* published in 1873, and new research in experimental physiology and pathology by Claude Bernard. Although

Hansen's discovery of the leprosy bacillus does not seem to have come to Johnston's attention, in his reports on the trials published in 1872 he believed that leprosy was caused by a specific '*contagium vivum*' or 'leprosy-germ', transmitted by inheritance and infection. He anticipated that the 'leprosy-germ' could be destroyed by boiling a solution of carbolic acid in water and applying the carbolic vapour to the skin and, by inhalation, to the pulmonary mucous membrane. The possible benefit of an improved diet given in conjunction with fumigation treatment was tested on a few selected patients.[26]

The practical limitations of British medical and governmental involvement in leprosy treatment in India are clear from Johnston's neglect of government-endorsed treatment trials in preference for fumigation. At the Palliport Lazaretto, all patients with the exception of one 'hopeless' case and four receiving cashew nut oil treatment were undergoing fumigation with carbolic vapour. Only two were following the gurjon oil trial requested by the Government of India.[27]

Indigenous remedies

Further, the promotion of cashew nut and gurjon oil treatments did not mean that local medical officers discarded indigenous remedies. Although Bhau Dajie's chaulmugra oil treatment had no government support, the Madras Leper Hospital reintroduced chaulmugra oil as a treatment in June 1874, a few months after the official, government-requested trial of Dougall's gurjon oil treatment had begun. In addition, when the Madras Leper Hospital's supply of chaulmugra was exhausted, marotty oil, often used interchangeably with chaulmugra as a leprosy treatment in indigenous medicine, was given an independent trial by Surgeon W. Macrae in 1875, using Dougall's methods. The results of the two-month trial were inconclusive, bringing Macrae to recommend to Drs Lewis and Cunningham, in their 1875 enquiry into leprosy for the sanitary department, that a longer trial be implemented.[28]

The extent, and thus the validity in scientific terms, of treatment trials proposed by the more distant government authorities could easily be curtailed by the interests of the regional medical authority. Local autonomy and indigenous medicine were not entirely supplanted by government initiative in leprosy treatments. British efforts to direct scientific medical investigation in India were ultimately dependent on the co-operation of the local medical service and presidency authorities. British scientific medical control in India was, at best, limited and susceptible to local impediment both from the leprosy sufferer, who was the subject of enquiry and experiment, and from the orientation and interests of the presidency medical authorities and local IMS officers themselves. Government-endorsed remedies for leprosy developed by the British neither inevitably displaced indigenous remedies, nor were free from competition from other remedies supported by British medical officers directly involved in leprosy care. The Madras medical officers were too distanced from central government authorities to experience the type of close struggle for control fought between the Royal College of Physicians and the British government in the early 1870s. However, their actions were as important in influencing the configuration of colonial powers as the political struggle of their superiors.

Disease theory and sanitary politics

Although gurjon oil was still being tested in 1874 when the leprosy bacillus was discovered, there was still no reliable cure for leprosy. Indigenous remedies such as chaulmugra and marotty oil remained local rather than international treatments and offered little hope. While there was no convincing cure, the discovery of the contagiousness of leprosy in 1874 raised the issue of containment and confinement of leprosy sufferers as a means of managing the disease and provided a further context for conflict over who would control leprosy management in India. By 1875 the Royal College of Physicians and the British government were no longer directing investigation of leprosy in India. Rather, leprosy investigation was being initiated by the Government of India and the sanitary department. In India, H.V. Carter of the Bombay IMS, supported by the secretary of state, rose to prominence as the authority on leprosy in Norway which continued to set the agenda for leprosy control. Inspired by the Norwegian experiment, Carter advocated similar legal and institutional measures to control leprosy in India. The aspirations of the sanitary department, however, saw Carter's sudden eclipse and the brief ascendancy of arguments against confinement of leprosy sufferers supported by J.M. Cuningham, sanitary commissioner with the Government of India.

Arguments for the isolation and segregation of leprosy sufferers

The Norwegian experience

In Norway, Hansen's discovery in 1874 that leprosy was caused by a bacillus, and the belief that leprosy could be transmitted by contagion, altered the leprosy control program already in effect. In 1856 a national registry of leprosy sufferers had been initiated in Norway in an effort to relate epidemiological information to public health measures and thus discover the cause of the disease. A secondary aim of the registry was to prevent the marriage of those with leprosy, though the prohibition of marriage was never actually implemented. A law passed in 1857 provided good hospital accommodation in order to encourage leprosy sufferers to hospitalize themselves. Indigent leprosy sufferers could, however, be forcibly confined. In 1875, on the basis of his findings on the contagiousness of leprosy, Hansen suggested that the existing opportunity for leprosy sufferers to be confined voluntarily be altered to enable the compulsory confinement of those with the most contagious form of the disease. In 1877 the Act for the Maintenance of Poor Lepers was passed, the first law in Norway to enforce hospitalization. This Act stated that those leprosy sufferers unable to work and thereby unable to support themselves would be hospitalized. Previously, the majority of those confined had either volunteered for hospitalization or sought refuge because of their poverty. It was expected that, with the support of these legal measures, the incidence of leprosy would significantly decrease in Norway.[29]

The 1877 Act was superseded by an Act of 1885 which authorized compulsory confinement of those leprosy sufferers living at home who refused to take measures against infecting other people. Unlike the 1877 Act, which applied only to the indigent leprosy sufferers, the Act of 1885 stimulated intense debate as it enabled the removal from their home of leprosy sufferers in better socioeconomic circumstances. The Bill was pushed through under pressure from Dr Hansen, supported by the growing conviction that leprosy was transmitted not only by inheritance but also, at least in part, by contagion. However, the concern in Norway not to 'interfere with the liberty of the subject' brought strong resistance to the Bill.[30] Opponents of the Bill argued that enforced isolation should not be imposed on those who were already suffering enough from their disease and that, since leprosy was already on the decline, there was no need for stricter measures. Although the theory of contagion was gaining acceptance in Norway, it was assumed, quite correctly, that close contact with a leprosy sufferer over a long period of time was necessary for transmission of the disease. Even after the passage of the 1885 law, lack of alarm about the contagiousness of leprosy, and a strong tradition in Norway of concern for patients' rights, meant that isolation measures were never complete. Hospitalized leprosy sufferers retained full freedom of movement and could even trade on the street, selling food and other items, though they were required to return to the hospital each night.[31] In India, the same issue of the requirements of disease control versus the liberty of the leprosy sufferer was to be raised by debates over legislation to control leprosy. However, fear of an Indian backlash was to prevent the law from ever permitting compulsory removal of leprosy sufferers from their homes.

In 1873, the Duke of Argyll, the secretary of state, deputed H.V. Carter to visit Norway in order to study the history and management of leprosy there. Carter had already made his mark with his article on the morbid anatomy of leprosy. He was committed to a high standard of medical research and eager to take up opportunities for career advancement. He had written to the *Lancet* criticizing the Royal College of Physicians' 1864 leprosy investigation for expecting satisfactory scientific information from medical officers who lacked a specific interest in leprosy.[32] Carter's first report on leprosy in Norway, published in 1874, recommended the intervention of government in the voluntary segregation of leprosy sufferers in India in the hope that, on the evidence of Norway, where voluntary segregation was practised, the disease might be eradicated. While Carter believed that leprosy was inherited, he was also impressed by Hansen's theory of contagion. On the basis that leprosy may be transmitted by both methods, Carter advocated both the isolation of leprosy sufferers from society and segregation of the sexes. He proposed the establishment of a network of asylums in India similar to that in Norway which he so admired, and suggested that in order to make isolation and segregation effective, legislation may be needed. This was a significant departure from the Royal College of Physicians' 1867 report which had declared leprosy to be noncontagious and argued against compulsory segregation. Even so, the Royal College of Physicians was prepared to support Carter in his hopes to investigate leprosy in India more fully.[33]

J.M. Cuningham and the anti-contagionist view

The British government's Army Sanitary Commission supported Carter, who was the secretary of state's man, endorsing his recommendations that a special enquiry be conducted into the pathological and sanitary significance of leprosy in India. Although not directly supporting Carter's recommendation for the building of asylums to confine leprosy sufferers, the commission did recognize Carter's view that leprosy was inherited. As the first step towards a full enquiry into leprosy in India, the Sanitary Commission approved the collection of accurate statistics on the numbers of leprosy sufferers in each locality.[34]

The Indian government was, however, far stronger in its criticism of Carter. In his notes on Carter's first report on leprosy in Norway, the Government of India's sanitary commissioner, J.M. Cuningham counteracted Carter's opinion that the reduction of leprosy in Norway was due to the segregation of leprosy sufferers, arguing that, on the basis of Carter's own evidence, segregation was only very partial in Norway. Segregation, Cuningham reasoned, was only an attractive theory of leprosy control for those who believed leprosy to be contagious or propagated by marriage, a belief for which, in Cuningham's opinion, there was insufficient evidence. Further, it seemed to him, the application of segregation in India would be an impossible task. The evidence that suggested segregation could eradicate leprosy was not sufficient to warrant the enormous cost of building asylums for all the leprosy sufferers in India. Cuningham still hoped that gurjon oil, tested after Carter's report had been formulated, would make segregation and such measures unnecessary. The only aspect of Carter's report on leprosy in Norway which Cuningham was able to endorse without reservation was the importance of gathering statistics on leprosy in India.

Cuningham, however, objected to Carter's proposal that a special leprosy commissioner be appointed to enquire into leprosy in India. Rather, Cuningham argued, Drs T.R. Lewis of the Army Medical Department and D.D. Cunningham of the IMS, who had been appointed as his specific assistants in 1874 to test the 'germ' theory of cholera transmission, should take up the microscopic and scientific aspects of the enquiry. Cuningham considered that the general facts of the enquiry could be dealt with properly only by the sanitary commissioner assisted by the civil surgeons acting in their capacity as health officers. Reflecting the influence of the sanitary department, the Government of India fully endorsed Cuningham's recommendations in its Resolution of March 1875 and it was decided that the next enquiry into leprosy in India be 'a sanitary enquiry, requiring the combined action of the whole sanitary department in India'.[35]

Indian Medical Service and sanitary department rivalry

Cuningham's resistance to Carter's proposals, particularly for the appointment of a special leprosy commissioner, was not simply a consequence of optimism at the effectiveness of the gurjon oil cure. Rather, it was consistent with the empire-building which had characterized the sanitary department from the appointment of the first sanitary commissioner in 1866. Cuningham's concern to control the gathering of information on leprosy was an aspect of the department's broader interest in carving out a professional status. Much as the Royal College of Physicians in Britain had used the investigation of leprosy to help develop its professional profile in the 1860s, the sanitary department employed control over leprosy research in its effort to extend, as far as possible, its authority over public health in India. At a time when the sanitary department was already embroiled in controversy over the management of cholera, with Cuningham facing criticism both from the IMS and from within the ranks of the sanitary department on his anti-contagionist stance, management of leprosy research offered an opportunity to assert the department's wavering power.

During the 1870s, a close rivalry developed between the IMS and the sanitary department. In the IMS, career prospects were poor, and rapid promotion of relatively low-ranking officers into positions of power in the sanitary department aroused fears that the IMS would become a 'second-class' service. Part of the IMS' frustration was the alleged 'specialist' status of members of the sanitary department. As with the Royal College of Physicians in London, the attainment of professional credibility for the IMS had been a struggle. Specialist appointments in the sanitary department, compounded by the comparative underfunding of the IMS, which made medical research virtually impossible, threatened the fragile professional status of the IMS officers.[36]

The sanitary department did little to reassure the IMS, being intent on asserting a professional position in relation to the existing Indian medical establishment. An editorial in the *Lancet* of 1869 noted that from the appointment of the first sanitary commissioner, 'no efforts have been spared to extend its sphere of action and increase its importance'. The principal means of achieving this end was the absorption of the functions and powers of other bodies. The commissioners were characterized by talent and ambition. Their ambition was to develop the Sanitary Commission into a public health department with centralized control over statistical information for the whole population, both Indian and European, and power to exercise sanitary surveillance over an area almost equivalent to the territory covered by the statistics. Following the events of 1857, the Government of India's anxiety about the health of its troops stimulated its support of a centralizing of sanitary statistics by the commissioners, in particular Cuningham. Cuningham sought direct control of mortality statistics, bypassing both civil surgeons and the provincial sanitary commissioner, much to the chagrin of the medical service which opposed such a reduction in their authority. By 1869 the authority of the Sanitary Commission was such that every matter sent for comment to the inspector-general of hospitals of the British or Indian Medical Services was sent to the sanitary commissioner for opinion and report before the Government of India would act upon it. In the opinion of the *Lancet*, the medical service, not the sanitary department should be medical advisers to the government as were the Royal College of Physicians in England.[37] However in India it was the sanitary department which increasingly played the role of the Royal College of Physicians.

Although Carter was strongly supported by both the secretary of state for India and the Army Sanitary Commission in Britain, the Government of India, supported and encouraged by the sanitary department, became increasingly irritated with Carter's requests for time and resources to continue his research into leprosy in India and other parts of the world. E.C. Bayley, secretary to the Government of India commented in 1876:

> Dr Carter seems to be getting rather beyond a joke. So far as I understand his researches have hitherto been utterly fruitless and useless to science. I am disposed to hint that the Government of India thinks that Dr Carter's special deputations should now cease, otherwise he will go on being deputed and writing accounts of his deputations for ever to no reasonable use.[38]

Carter had little backing from Cuningham, who had refused to support him in a dispute with Gavin Milroy, author of the *Report on Leprosy and Yaws in the West Indies*, published in 1873, and endorsed by the Royal College of Physicians as the investigator of Beauperthy's treatment. Carter claimed that Milroy had 'deprived him of the merit which, he ... deserves for shewing the importance of segregating lepers and for providing asylums for this purpose'.[39] Cuningham was increasingly sharp in his criticisms of Carter. He advised against the Government of India publishing and thereby giving official sanction to Carter's findings on the possibility of a *micrococcus* being a specific cause of leprosy. Cuningham argued that European authorities disagreed with Carter's research and, more significantly for Cuningham's efforts to bolster the position of the sanitary department, that Drs Lewis and Cunningham considered Carter's findings to offer nothing new.[40] Hansen's findings were, however, to prove Carter right.

Carter's abilities in scientific research were beyond doubt. In addition to pursuing ground-breaking research into leprosy, Carter had discovered the fungal nature of the condition known as 'Madura foot', had published on relapsing fever, leprosy and elephantiasis, had pursued various researches into microscopy, and had published research on cutaneous diseases in conjunction with the eminent doctors Fox and Farquhar.[41] However, Cuningham was not above destroying the career of an opponent. During the same period, he had made every effort to silence Surgeon-Major A.C.C. De Renzy, sanitary commissioner for the Punjab, who had argued against Cuningham and the Government of India's anti-contagionist stance on cholera. According to the *Lancet*, it was not Carter's ability but the position of the Government of India in opposing the segregation of leprosy sufferers in asylums, encouraged by the sanitary department, which caused the appointment of Lewis and Cunningham to pursue an Indian government-endorsed investigation into leprosy in 1875, 'elbowing Dr Carter out of any participation in a matter upon which he has thrown great light by honest and admirable work'.[42]

At the time of the announcement of Lewis and Cunningham's enquiry, the *Indian Medical Gazette*, which claimed to represent medical opinion in India, was expressing what would in the modern day be called 'enquiry fatigue'. Scepticism as to the value of past medical enquiries into leprosy, including those conducted by Fox and Farquhar and the Royal College of Physicians, extended to include the sanitary department's new investigation. The sense of ennui was well founded. Despite the aggressiveness of the sanitary department, Lewis and

Cunningham's 1875 enquiry delivered little. The Government of India commented that in general the report offered little of value and added nothing new to the understanding of leprosy in India. Its principal contribution was in endorsing Cuningham and the Indian government's opposition to Carter's recommendations, specifically, the confinement of leprosy sufferers and the appointment of a special leprosy commissioner.[43]

Carter did continue to pursue leprosy research for the authorities of British India and persisted with the argument for segregation and confinement. His 'Memorandum on the Prevention of Leprosy by Segregation of the affected' was published by his presidency's government in the *Bombay Government Gazette* of 1884, and 'Further Observations on the Prevention of Leprosy by Segregation of the affected' followed in 1887. However, being opposed by the sanitary department, Carter never received the level of support from the Government of India which could have furthered his professional advance. In 1884, he was passed over for the position of provincial sanitary commissioner for Bombay. The retiring incumbent, T.G. Hewlett, was reappointed for a further four years, an unprecedented move which effectively curtailed any hope of further promotion for Carter. Carter was so frustrated and outraged by the injustice of this move that he sent a memorial to the secretary of state requesting that the appointment be disallowed, but with no effect.[44]

The sanitary department's efforts to limit both Carter's role in leprosy research and his influence over government, were effective. In the Home Resolution of 26 September 1888, the Government of India proposed that it play a more direct part in the prevention and treatment of leprosy. The resolution included Carter's opinion on segregation, noting that the governor-general in council 'is assured that no measure could effectually stamp out the disease which is stopped short of the absolute segregation of the sexes, and confinement for life of all affected by it'. However, the Government of India complied with sanitary department opinion opposing the implementation of such measures, since they 'would not only be repugnant to public opinion at the present time, but would in India be perfectly impracticable'. The Government of India resolved that it could do no more than encourage the leprosy sufferer to voluntarily seek medical and charitable relief in hospitals and asylums, enforce segregation of the sexes in state-funded asylums and encourage the same practice in those maintained by voluntary funds. There was no pressure from D.D. Cunningham to legislate to control leprosy. In his capacity as special assistant to the sanitary commissioner and thus chief scientific adviser to the Government of India, he strongly argued, in agreement with Cuningham, against the view that leprosy was contagious.[45]

Effectively, the 1888 resolution served only to endorse the existing practice of leprosy care and to recognize the enormity of the task of enclosing India's entire population of leprosy sufferers. Prior to drafting their resolution, the Government of India circulated local governments and administrations requesting information about the number of leprosy sufferers in each area and whether or not those in institutional care were segregated. The Madras government reported that of the 14 175 leprosy sufferers recorded in the Madras census, a figure considered by the Indian government to be under-reported, several hundred only were in institutional care, and that strict segregation of the sexes was enforced in each hospital by the medical officer in charge. In the Madras and Cochin asylums, men and women were locked into their separate wards at night and even husbands and wives were prevented from cohabiting. However, the understanding of the degree of segregation practised seemed to differ according to the proximity of the reporter to the hospital. In the following year, Surgeon-Major H.D. Cook, who was in charge of the Madras Leper Hospital, noted that complete segregation of the sexes did not exist in the asylums of the Madras presidency and that contact and conjugal relations between asylum patients and their spouses were common. The responses from other provinces were consistent with the Madras findings that only a very small proportion of leprosy sufferers in India actually took advantage of the institutional care available.[46]

Father Damien and European fear of leprosy

The Indian government's non-interventionist response to leprosy management in India was, however, short lived. The last illness and death of Fr Damien, who had contracted leprosy while caring for the leprosy sufferers of Molokai, produced an hysterical response in Britain, triggering public fear that leprosy was an imminent danger to Europeans who had seemed virtually invulnerable. At a time of active imperial expansion and world migration, when leprosy was identified increasingly with non-European races, particularly Chinese and African peoples, Fr Damien's death made clear the European susceptibility to leprosy. There was fear in Britain not only that leprosy would spread to Europe but also that by living in countries where leprosy existed, Europeans might contract the disease. Never had the vulnerability of the European to infection with leprosy been so publicly displayed and seemed so tragic. The previously moderate response to the notion that leprosy was contagious was transformed by fear. An indication of public alarm in Britain, was the dramatic increase in discussions of leprosy in *The Times* of London between 1889 and 1891.[47]

Several publications of the time both reflected and fuelled fear of infection. H.B. Wright, rector of Greatham, Hertfordshire, fearful of the rapid spread of the disease, mustered the arguments in favour of the contagiousness of leprosy, citing Carter, among others, in support. In 1889, Wright published *Leprosy an Imperial Danger*, prefaced with the prayer: 'May God preserve my country from leprosy'. Wright not only argued that 'leprosy comes always direct from the leper', but articulated the racism which was contributing to British public fear of the spread of leprosy to Europe. While recognizing that leprosy had previously existed in Europe, Wright identified the disease with specific non-European races, describing the Africans sold into slavery in America as a 'highly leprous' race and maintaining that Asia was 'possessed of inexhaustible sources of the malady'.[48] Further, in 1889 C.N. Macnamara reissued his 1886 publication, *Leprosy a Communicable Disease*, expecting that in the context of Fr Damien's death, his earlier researches might prove useful to those considering the subject. Macnamara's view that leprosy was, like syphilis, sexually transmitted, was, for the anxious British public, more than an endorsement of Danielssen and Boeck's hereditary theory of leprosy transmission. The association of leprosy and sexuality, which in medieval Europe had meant that the leprosy sufferer was attributed with lustfulness and voracious promiscuity, deepened the fear of racial contact and mixture already felt by the European colonizing nations.[49]

Changes in disease theory, Carter's research and the death of Fr Damien also had a substantial impact on public opinion in Madras regarding leprosy. In a letter to the Government of India, G. Bidie, surgeon-general with the Madras government, summarized the changed attitude evident in the presidency. Since the Royal College of Physicians' report of 1867 which had stated that leprosy was not contagious, Bidie noted:

additional information has been accumulating which has turned the current, not only of professional but also of public opinion, rather in the opposite direction. This and the melancholy fate and self abnegation of Father Damien have led [sic] so stimulated public interest in the matter, as to induce a tendency to panic on the subject of leprosy.[50]

The National Leprosy Fund and the Leprosy Commission

Fr Damien's death, however, did not only induce panic in the British Empire. It also inspired in Britain a sense of public responsibility for the leprosy sufferers of India. Fr Damien had become the embodiment and example of 'noble self-sacrifice'. To mark the Empire's sympathy with Damien's sacrifice and its sense of moral indebtedness, the National Leprosy Fund was established on 17 June 1889. The Fund was an organization composed of members of the aristocracy and eminent church and medical gentlemen of the British realm, united under the presidency of the Prince of Wales.

Opening the first meeting of the Committee on 17 June 1889, the Prince of Wales recalled Damien's 'heroic life and death', claiming that he had not only 'roused the sympathy of the United Kingdom' but also 'brought home to us that the circumstances of our vast Indian and Colonial Empire oblige us, in a measure at least, to follow his example'. The duty of the empire was not to 'foreigners and strangers, but for our own fellow-subjects'. Inspired by the nexus of Fr Damien's death with a burgeoning sense of imperial responsibility, the Prince of Wales proposed a three-fold scheme embracing the establishment of a memorial to Damien at Molokai, the formation of a fund to support indigent British leprosy sufferers in the United Kingdom, the endowment of two studentships for the study of leprosy and the appointment of a Leprosy Commission to investigate leprosy in India.

The Leprosy Commission, which began its investigations in November 1890, was appointed by the executive committee of the National Leprosy Fund in consultation with the established British medical profession. The idea of a Leprosy Commission investigation had been, in part, prompted by calls from the leprosy committee of the Royal College of Physicians for a fresh enquiry into leprosy and its contagiousness.[51] Reflecting the professional security of British medicine by the late nineteenth century, the Royal Colleges of Physicians and Surgeons both had sufficient kudos to be involved in directing a leprosy investigation under the patronage of the British royal family. The three British representatives were Mr Alfredo Antunes Kanthack, FRCS, of St Bartholomew's Hospital, London, appointed by the Royal College of Surgeons; Mr B.N. Rake, of Trinidad, appointed by the Royal College of Physicians; and Dr G. Buckmaster, appointed by the Committee of the National Leprosy Fund. In addition, two Government of India medical officers were deputed to assist the British appointees: Surgeon-Major A. Barclay, the secretary to the surgeon-general of India and Surgeon-Major S.J. Thomson, deputy sanitary commissioner, second circle, North Western Provinces and Oudh. Like the Royal College of Physicians' earlier enquiry, the Leprosy Commission was substantially dependent on local IMS expertise.[52]

The rise of the Indian Sanitary Department could not prevent the Royal College of Physicians' influence in directing British attention to leprosy in India from continuing to the close of the nineteenth century. Nor could it limit the significance of local medical authorities in the achievement of effective leprosy research. By the end of the 1880s, the British attitude to leprosy in India had significantly changed. Influenced by the arguments against the confinement and segregation of leprosy sufferers put forward by the sanitary department in the 1870s, the Government of India resolved in 1888 that legislation against leprosy was inappropriate and impracticable. By 1889, however, fear of contagion sparked by the death of Fr Damien had changed public opinion. Legislation to control leprosy was again on the agenda.

Colonial science, including colonial medical science, may have been perceived by the metropolis as a form of 'low science', identified only with fact-gathering and pursuing questions set by Europe as MacLeod has argued. However, in terms of leprosy research at least, the relationship between metropolitan and colonial medical science was more complex and less unequal than it appears from the metropolitan position. British medical treatment of leprosy in nineteenth-century south India supports Arnold and Harrison's arguments that a distinctive British medicine developed in India in this period, not dependent on Europe for direction, though connected with medical developments there.[53] While investigation of leprosy and publication of research was stimulated in the 1860s by Europe, by the 1870s India was taking the initiative. The actual treatment of leprosy pursued by the British in India did not slavishly follow European guidelines. The refusal of the Madras medical officers to trial Beauperthy's cure fully, the incorporation of the indigenous remedy chaulmugra oil into British medical practice, the adaptation of gurjon oil to leprosy treatment and the application of the latest European medical theory in Johnston's fumigation treatments, showed the capacity of the IMS to adapt indigenous remedies and apply European scientific and medical developments in a manner appropriate to Indian conditions.

The discovery of alleged cures substantially increased British and Indian government attention to leprosy. From June 1878, 100 copies of all reports relating to leprosy were required by the medical branch of the Home Department to meet the needs of the Colonial Office, instead of the 25 previously requested.[54] From the mid-1870s, government attention to leprosy in India was increasingly directed by the Indian government, and leprosy research became entangled in the power struggle between the Indian Medical Service and the sanitary department established in the 1860s. The failure of the leprosy cures proposed in the early 1870s contributed to a shift in attention away from cure and towards containment of the disease. Segregation and confinement became the critical issues in leprosy management driving debate towards the 1898 Lepers Act. In this instance, rather than being a tool of power over the colonized, medicine was used by the colonizers to dominate each other.

7
Confining Leprosy Sufferers: the Lepers Act

If the initiation of the Leprosy Commission was a reflection of both British admiration for Fr Damien and fear of the disease, then the Government of India's March 1889 decision to employ 'special legislation' was also as much a reflection of fear as of the change in viceroys in December 1888 from Dufferin to Lansdowne. On 15 June 1889, 'a Bill to make provision for the isolation of lepers and the amelioration of their condition' was circulated for comment to all local government and administrative authorities in India.[1]

The 1889 Leprosy Bill

The 1889 Bill revived the issue of forced confinement of vagrant leprosy sufferers raised in the 1840s, and again by the Royal College of Physicians' report. As before, it was not those of the higher socio-economic classes with leprosy, but those who were also vagrant and poor, the majority of whom were Indian and Eurasian, who were identified by British authorities as a group for whom, in Foucault's terms, confinement was a natural condition and the asylum a true 'homeland'.[2]

The limitation of confinement to vagrant leprosy sufferers alone and the absence of disease-control measures to regulate the work and, if necessary, movements of non-vagrant leprosy sufferers, were matters of dispute among British and Indian authorities, as they had been in Norway. Criticism of the 1889 Leprosy Bill marked the beginning of a decade's debate and negotiation over the terms of legal compulsory confinement of leprosy sufferers in India. Comment on the Bill from the Madras presidency indicated the extent of conflict among medical, legal and administrative authorities within the presidency, and between the presidency and Government of India and the secretary of state over what precisely were the requirements of disease control, and whether they should take precedence over Indian public opinion.

As occurred in the context of the 1896–1900 plague, it was the Indian educated and professional classes who assumed the role of the articulators of Indian public opinion in Madras. Their involvement was not unwelcome. Indeed, the Madras government was unique among the presidency governments in appointing a committee to represent to it Indian opinion on the Bill.[3] Rather than following 'public opinion' as expressed through the press, the Madras government communicated directly with the formulators of 'public opinion'. The calibre of those members of the Indian community offering their views reflected the importance of the 1889 Bill. P. Anandacharlu had headed the Ripon Reception Committee, and M. Viraraghavachariar was managing proprietor of the *Hindu*, which provided a context for the emerging political voice of the Madras professional elite. Both were closely involved in Madras politics and were founding members of the Madras Mahajana Sabha, an organization committed to raising south Indian political consciousness and inspiring participation in decisions affecting India's political future. Formed in 1884, the Madras Mahajana Sabha was a unified and powerful alliance of the Madras professional Indian elite broadly representing Indian upper-middle class opinion, including that of the merchant and landlord class. It included the majority of south Indian local associations and literary societies, and was affiliated with the numerous political bodies which had developed in the presidency by the 1880s.

K.P. Sankara Menon, as an educated professional, was typical of the non-brahmin men who joined the Brahmin-educated elite in leading the formation of Indian public opinion on the 1889 Bill. He held both BA and BL degrees, was a vakil of the Madras High Court and an office holder in the Madras Mahajana Sabha in 1886–7.[4] In 1889 he was provisional secretary of the newly formed Madras High Court Vakil's Association, which included P. Anandacharlu on its Managing Committee from 1889–93.[5] While female opinion on the leprosy legislation no doubt existed, as with much social debate in nineteenth-century colonial contexts, such views tended to be privately expressed and were not ostensibly part of the public debate. Unfortunately, the opinion of Indian women on legislation to control leprosy has proved inaccessible.

The Madras Indian upper-middle-class authorities endorsed forced confinement of vagrant leprosy sufferers but insisted on the protection of all those in employment from such an indignity. Like the discussions of the first half of the nineteenth century, the passage of the Lepers Act was to prove indicative of the disunity among those in power over the extent and terms of confinement. The 1898 Lepers Act was not the utterance of a single colonial authority but represented a negotiated position.

Responses from the Madras presidency

The principal point of dispute over the terms of the 1889 Bill was its emphasis on the confinement of vagrant leprosy sufferers to the detriment of measures specifically to control those who might spread the disease in the home or through work. In contrast with the early-nineteenth-century discussions, in the debates on the 1889 Bill it was not the Madras medical authorities who strongly advocated confinement of vagrant leprosy sufferers but the Government of India. Although not unanimous, the comments forwarded to the Government of India by British medical administrative and judicial Madras authorities and endorsed by the Madras government showed a resistance to the use of the Bill to control vagrancy without including measures to combat the spread of disease by the non-vagrant population. G. MacWatters, the collector of Salem, commented that whereas the Bill

> provides for what may be called voluntary lepers – those who apply of their own accord for admission to a retreat (section 4); and for vagabond lepers – those found wandering without any ostensible means of living (section 5) . . . no provision is made for the non-voluntary and non-vagabond lepers.

He noted that the Government of India was probably 'aware of the omission' but not willing to take 'too drastic a measure to start with' and legislate for the 'removal of members of respectable families from their homes against their consent'.[6] Indeed, this was precisely the grounds of the Government of India's caution in legislating to control leprosy in 1888.[7]

The 1889 Bill sat uneasily between two types of legislation passed by the Government of India from the mid-nineteenth century: disease-control legislation which inevitably involved some measures to control the disease-sufferer, and legislation which dealt solely with the control and confinement of people who were perceived as a public nuisance. Government of India legislation to manage venereal disease had been passed in 1868; however, it had been repealed in 1888 as ineffective, expensive, morally unacceptable for its apparent legitimization of prostitution, and of less importance in the decision to repeal the Act, because it was contrary to Indian public opinion. Legislation to control smallpox was passed in the 1870s and 1880s. In each instance, whatever their social consequences, these laws were primarily measures to protect the colonizers from disease.[8]

Its focus on confinement of poor and vagrant leprosy sufferers allied the 1889 Bill more closely with later nineteenth-century attempts in Britain and India to legislate against public drunkenness and vagrancy than with laws ostensibly for disease control. Section 4 of the 1889 Leprosy Bill, which provided for the voluntary confinement of leprosy sufferers in designated retreats, was based on the sections of the British Habitual Drunkards Act of 1879 which allowed for the confinement of 'habitual drunkards' at their own request.[9] More significantly, the 1889 leprosy Bill also closely resembled the Government of India's legislation against European vagrancy. The European Vagrancy Act of 1869 had been amended in 1871 and both Acts repealed by the India Act IX of 1874. The 1874 Vagrancy Act applied only to persons of European extraction born in European, American or British colonies and their male descendants, and specifically excluded Eurasians and Indians. This is in contrast with the 1889 Leprosy Bill, which, although Europeans were subject to its provisions, primarily concerned Indians and Eurasians, who were the majority of leprosy sufferers. There was no equivalent legislation to control Eurasian or Indian vagrants. Eurasian and Indian indigent people usually came within the provisions of the municipal bye-laws. The provisions of the 1874 Act concerning the powers of police to take up any vagrant European found 'without employment or visible means of subsistence' were closely paralleled in the 1889 Leprosy Bill. On confirmation that the person detained was a vagrant they were then confined in a workhouse. Similarly, on confirmation of leprosy, the leprosy sufferer was to be detained in a leprosy retreat.[10]

Indicating the concern in the Madras presidency that the Bill would be used arbitrarily as a vagrancy law, and emphasizing the need to incorporate measures more specifically aimed at disease control, the first section of the Bill to be targeted for criticism by the Madras authorities was the Government of India's definition of the term 'leper' as a person with respect to whom a certificate that he is suffering from leprosy has been made by a medical practitioner having from the Local Government general or special authority, by name or in virtue of his office, to certify as to the existence or non-existence of the disease in any person alleged to be suffering therefrom.[11]

This definition of the 'leper' allowed much opportunity for error and harassment. F.A. Nicholson, collector of Tinnevelly, was well supported in his concern that the Bill should include 'a proper definition of that leprosy which will warrant a magistrate in taking action', since 'there are thousands of lepers who, except in a potential danger to offspring, cannot be said to be dangerous to society, since they have not arrived at the really dangerous stages, such as the ulcerated tubercular form'. The 'native gentlemen' consulted perceived a potential for social harassment without a proper definition of leprosy to protect 'ignorant sufferers from seeming leprosy, from the petty annoyances of arrest'. The surgeon-general for Madras, G. Bidie, cautioned that: 'the authority of deciding as to whether or not a person is a leper . . . will require to be carefully guarded' and proposed that 'as far as possible, [it] be restricted to commissioned officers in independent charges'.[12] Local authorities were the first line of defence for the vagrant leprosy sufferer.

The 'leper retreat': hospital or prison?

While at the local level every effort was made to move the Bill towards disease control and away from control of the vagrant poor, the role of the 'leper retreat' also came into question. By contrast with the advocacy of greater restraints on leprosy by Lawder in 1840, in 1889 it was the presidency government which suggested that the 'leper retreat' should be more a hospital than a prison. Under the Bill, the 'retreat' was simply defined as 'a place for the time being approved by the local government as suitable for the accommodation of lepers'.[13] 'Leper retreats' were to be funded with money earmarked for hospitals and medical institutions. Bidie commented that to use funds allocated for medical relief for their support 'would be hardly fair or logical if a retreat is to be simply a sort of prison'. Acknowledging that leprosy was then 'practically an incurable disease', Bidie recommended that the retreat be made less a prison and more a hospital where at least 'to relieve the sufferings of the leper and prolong his days . . . skilled medical aid would be available'.

He predicted that a shift in the retreat's role away from confinement and towards care would be 'a strong inducement to many sufferers to enter', whereas 'merely secure seclusion. . . . will hardly satisfy the public and may prove an obstacle to lepers entering such places'.[14]

For those designated 'vagrant lepers', the retreat was indeed a prison-like institution. The Government of India claimed presence of disease to be the basis for identification as a 'leper'. However, as in the early nineteenth century, only where poverty and vagrancy co-existed with leprosy did identification as a 'leper' became a condition requiring confinement. The option of choosing confinement was reserved for those leprosy sufferers possessing 'employment or visible means of subsistence'. According to the Bill, 'Any person knowing or believing himself to be suffering from leprosy who desires to be admitted into a retreat may apply orally or in writing to any Magistrate for admission thereto and for detention therein either for life or a term of years.'[15] The status of the 'leper retreat' was for the higher classes ambiguous, but for the vagrant it signified permanent confinement.

Disease control versus vagrancy control

Medical, legal and judicial authorities in the Madras presidency agreed that confinement of vagrants was only valid where a real threat to the public from disease existed. They considered that exemption from forcible confinement for those leprosy sufferers of a better socio-economic position was a form of class discrimination which hindered the management of leprosy. As Spring Branson noted:

> The danger of contagion by personal contact is much greater in the case of lepers having a 'visible means of subsistence' than in that of leprous beggars and vagrants. The danger arising from the sexes not being segregated is common to both classes, but is likely to be greater in the case of those having sufficient means to marry into healthy families and to stay the destructive effect of the disease for a time by medical aid, care and nourishment.[16]

Local Indian authorities agreed that measures against vagrants with leprosy, though not entirely unwelcome as means of vagrancy control, were of very little benefit in the eradication of the disease. The eminent Mr Justice Muttusami Aiyar, the first Indian to be appointed *sub-protem* judge to the Madras High Court, argued that 'compul-

sory detention in retreats of vagrant lepers will only tend to remove a public nuisance'.[17]

Protection of class

While recognizing the inadequacy of a law focused on vagrants
suffering from leprosy, the British authorities in Madras acknowledged that imposition of compulsory confinement on well-to-do
leprosy sufferers would be perceived as an intolerable interference
by the Indian upper classes. Further, Nicholson, the collector of
Tinnevelly, noted that even voluntary isolation of leprosy sufferers
who still had family and livelihood was contrary to Indian culture:
'The extraordinary domesticity of the Hindu, and the affectionate
carelessness of friends and relatives, will be an absolute bar to any
voluntary seclusion in a retreat by any who have even the sem-
blance of a home or means of maintenance.' LeFanu, collector of
North Arcot, reporting on the Indian response to the Bill, noted:
'The consensus of opinion is that there is no objection to the Bill;
and that is goes [sic] exactly as far as it is safe to go and no further'.
Even though he considered it could be desirable to extend the Bill's
frame of reference, 'in the present state of public opinion', he con-
ceded, it was 'impolitic to extend the measure further so as to bring
lepers of all classes within its provisions'. Only W.A. Willock, collector of Vizagapatnam, despite recognizing the public objection to
interference in family life, was bold enough to suggest using legislation to introduce an ambitious 'system of registration of all lepers'
in areas where the disease was most prevalent,[18] a practice which
had been part of Norway's leprosy control programme since 1856.

It was not only the leprosy sufferers living with their families
who were protected by Indian middle-class public opinion from
forcible confinement, but also those who lived by their own labour.
Such people formed a significant proportion of the publicly visible
population with leprosy. Nicholson reported that in Tinnevelly alone,

> in numerous cases lepers sell eatables, such as buttermilk, veg-
> etables, betel, etc.; one is known to keep a public eating-house;
> another distributed food in a temple (!); one well-to-do leper hired,
> by gifts of food, a little boy from a healthy family, to scratch him;
> the boy contracted the disease at first in his right forefinger.[19]

The Tamil press indicated that Indian public opinion was not, however, entirely in support of such activities by leprosy sufferers. In

1874, the Tamil paper, *Dinavartamani* observed that leprosy sufferers mixed freely with the public and sold milk and vegetables at the Chintadripettah bazaar, objecting that: 'This state of things should not be allowed when there is a Leper Hospital'.[20]

Even though leprosy sufferers in employment were regarded by almost every local Madras authority as the greatest source of contagion, they were completely omitted from the 1889 Bill. Both British and Indian authorities who were consulted strongly recommended the implementation of measures to prevent those with leprosy selling food and drink and taking domestic work or other similar employment. However, their compulsory confinement was never advocated by the Indian commentators, and rarely by the British, who feared a backlash from the Indian middle classes. Thus, when compulsory confinement was proposed by the British it was as a last resort, only to be applied if the imposition of fines or confiscation of the leprosy sufferer's goods for sale or tools of trade had been insufficient deterrents.[21] It was only Bidie who suggested adding a clause to the Bill which made it a penal offence for leprosy sufferers to engage personally in any employment associated with providing food, water or clothing to the public, and prohibited their mixing in public assembly, transacting business in public office and seeking any kind of service 'without declaring themselves to be suffering from leprosy'.[22] The Lepers Act of 1898 was not a simple reaffirmation of the 1889 Bill but a result of negotiation between competing priorities: control of vagrancy and control of disease.

Protection of liberty

The British legislators assumed that vagrant leprosy sufferers possessed less claim to liberty than those of the higher social classes though this was not the view of the sufferers themselves. The Government of India was prepared to act against vagrant leprosy sufferers despite being aware that: 'The great majority of lepers prefer to be objects of local charity and seem to regard with disapproval even the small amounts of restraints to which they are liable in institutions maintained for their treatment.'[23] Perception of a distinction between the rights to freedom of the vagrant and the better-off classes was not unique to India. The same distinction was apparent in the passage of legislation to confine leprosy sufferers in Norway. In Norway, the 1877 Act to confine indigent leprosy sufferers was passed without struggle. However, the legislation passed in 1885 to enable the removal of those with leprosy from their homes, first had to overcome the reluctance of parliament to 'interfere with the liberty of the subject'.[24]

Given the existence, since 1869, of legislation controlling European vagrants in India, the distinction cannot be construed as simply a racist effort by colonial British authorities to exert superiority over the predominantly Indian population of vagrant leprosy sufferers. It was more a matter of class than race, both British and Indian upper classes sharing the same view. Prominent Indian citizens in Madras assumed that vagrant leprosy sufferers had less claim to liberty than those with leprosy of the higher social classes. P. Anandacharlu and M. Viraghavachariar approved of the Bill on the grounds that it 'does not unduly interfere with individual liberty and affords a salutary check to the spread of a loathesome disease'. The reminder from the collector of Vizagapatam that '[t]he pauper lepers themselves would prefer their present mode of life to the greater comfort of the retreat combined with the loss of liberty' was given little weight.[25]

The Indian upper classes perceived confinement as more 'natural' and acceptable for the vagrant leprosy sufferer, signalling to the Government of India that any action taken against the 'pauper leper' would not provoke an Indian backlash. Such reassurance was particularly necessary in 1889 when controversy over the Age of Consent Bill raised the thorny issue of how much, if any, state intervention would be tolerated in matters affecting Indian society. The British in India had learnt well the lessons of 1857 and were reluctant to introduce any measures which could threaten British authority, particularly at a time of burgeoning nationalist consciousness.

Liberty was perceived by the Indian middle classes as an attribute of the higher socio-economic classes which deserved and required protection. Those leprosy sufferers who were employed were regarded as immune from confinement even though their work was recognized as more likely to transmit leprosy than earning a living by begging. This class distinction may be, in part, attributed to emulation of British values by the English-educated Indian elite, but was also entirely consistent with traditional Indian attitudes to leprosy. The sufferer who could continue to work remained part of Indian public and domestic life. It was when work was no longer physically possible that he or she became 'uncared for by all'.[26]

Indian middle-class efforts to protect the liberty of those in employment also suggest a concern to protect the Indian community from undue and unwelcome British interference, reflecting the climate of increased political awareness of the neglect of Indian rights under British rule. While recommending that the Bill should be amended to restrict the activities of those leprosy sufferers in employment who posed a medical threat, the Madras Mahajana Sabha made every effort to suggest restraints on leprosy sufferers in employment which 'will least interfere with their freedom of action'. Rather than confinement, the organization merely advocated a fine to 'prevent lepers from following the occupation of vendors of articles, of dress or of food and drink [sic]'.[27] It is likely that the Sabha's reluctance to confine leprosy sufferers in employment was also a reflection of the substantial influence of middle-class merchants within the Sabha[28] and their objection to legal interference in trade and mercantile activities in Madras.

Confinement and criminality

In the 1889 Bill careful attention was paid to class in determining who could be legally confined. However, once the leprosy sufferer was institutionalized, irrespective of whether he or she was well-to-do or a vagrant before admission, the provisions of the Bill ceased to recognize class distinctions. Vagrant and poor leprosy sufferers did enter hospitals of their own volition. However, in the discussions on the 1889 Leprosy Bill it was generally assumed that voluntary patients in leprosy hospitals were not vagrant and poor but were of a higher socio-economic status. The enforcement of confinement was perceived by the Government of India, and even by senior Madras medical authorities, to be appropriate to the leprosy sufferer of any class once in the retreat. According to Section 7 of the 1889 Bill, both the arrested and voluntary patients were liable to be arrested and returned to the retreat if they left without being formally discharged. Implicit in the arrest of pauper vagrants was the assumption that poverty was a criminal condition. Not only poverty but confinement itself could confer the attribute of criminality on those with leprosy.

The contrast between Bidie's description of the leprosy sufferer who lives outside the retreat and his recommendations for the treatment of those inside, is strong evidence for a transformation in the colonial perception of the leprosy sufferer once the sufferer was confined. Bidie described the leprosy sufferer of any class who lives outside as possessing options and preferences. He noted that

> it is not likely that well-to-do lepers will thus seek to immure themselves in a simple retreat, as no personal advantage will be likely to accrue from mere seclusion, and as it will entail separation from friends and association with a number of persons in all stages of a loathesome disease. As regards the poor, some of them may enter for a bit of bread and shelter, but ordinarily they will prefer staying outside in the enjoyment of freedom and of the alms which their disgusting condition enables them as beggars to extort.[29]

However, although Bidie had argued against the use of a leper retreat as a prison, he did perceive the confined leprosy sufferer as a prisoner. He suggested that in order to deter escape, a penalty, such as 'solitary confinement on re-capture', should be imposed on 'fugitive lepers', otherwise, he argued, 'the inmates, – unless a very vigorous watch is kept, will walk out and in as they please'. Rather than leaving punishment of misconduct or escape to the discretion of the local government as stipulated in the 1889 Bill, Bidie recommended that the methods of punishment should be enshrined in the Bill at Government of India level. Bidie cited punishments exercised by superiors of the old European leprosy asylums as appropriate for the control of independent-minded leprosy sufferers. Although not going so far as to recommend the presence of 'a gallows kept ready at the end of the house', as in the Edinburgh Asylum, he reminded the chief secretary to the Madras government that in the old hospitals: 'The Dean, Prior, or other superior... had the power of punishing the offenders with the birch, by putting them on bread and water diet, [or] ... ejection....'[30]

His opinion was clear, that once confined, the leprosy sufferer lost the privileges of freedom which a hospital patient would enjoy and became an inmate of virtual criminal status subject to punishment. The suggestion that physical punishment should be applied to those with leprosy was a significant departure from early-nineteenth-century thinking in which even physical labour was considered by the Madras police to be inappropriate for leprosy sufferers.

The Government of India Bill paid careful attention to class difference in determining who could be confined legally. However, once the leprosy sufferer was confined, the provisions of the Bill ceased to recognize class distinctions and conferred the same penal status on the voluntary resident of a 'leper retreat' who might be middle class, as on the involuntary resident who was of the lowest socio-economic status. Within the retreat, confinement became the natural state of all those with leprosy, irrespective of class, and the potential of the retreat to be a type of hospital became subordinate to its role as a penal institution.

Representatives of the Indian upper-middle classes strongly dissented from British opinion on this matter. Sivangnana Mudalier, writing on behalf of the Trichinopoly Municipal Council, objected that the Bill 'treats all escaped lepers alike and makes no distinction between the leper detained with his free will and consent and the leper arrested and locked up'. He argued that voluntary residents of the asylum ought not to be returned if they leave, and should be free to come and go as they wished. However, he continued, the involuntary resident, who enters the retreat 'under arrest and order of the District Magistrate . . . when he escapes . . . is guilty of an offence, which the Council think should be made punishable'.[31] The Indian authorities sought to sharpen the distinction between the voluntary and involuntary resident by asserting that the involuntary resident's status was essentially criminal. When the forcibly confined leprosy sufferer escapes he is 'guilty of an offence' and therefore 'punishable', but when the voluntary resident follows, he is perceived as having done no wrong. The Indian authorities vehemently opposed the attribution of a penal character to the well-to-do who sought treatment for leprosy within the retreat. They sought to ensure that, even if it were a prison for the involuntary residents, the leprosy retreat remained a hospital for the voluntary.[32] Although vagrant leprosy sufferers often did voluntarily seek residence in leprosy hospitals, in the Indian discussion of the Leprosy Bills, voluntary residents were exclusively identified as being of higher socio-economic status.

Indian middle-class consciousness

Indian upper-middle-class responses to the 1889 Bill indicated a strong sense of class distinction within the Indian community. Those of the lowest socio-economic level were perceived as having the least claim to liberty, those who were employed in menial, domestic or shop work as having greater claims and those who had sufficient wealth and status to remain in the home, away from public view, the greatest. The 1889 Bill provided a context not only for the expression of Indian professional opinion on the merits of disease control but also for the affirmation of Indian class distinctions. The strong objection among the Indian upper-middle classes to the blurring of caste and class lines within the asylum was consistent with the Indian middle-class response to British strategies against plague. As Arnold notes, the strength of Indian protests against plague measures such as 'house searches, compulsory hospitalisation and the inspection of rail passengers was precisely that these measures treated Indians as an undifferentiated whole, without acknowledging differences of caste and class'.[33]

In expressing a clear sense of the difference between the vagrant poor and the employed and well-to-do members of the Indian population, and in ascribing criminality to the vagrant Indian poor, the Indian upper-middle-class response was similar to that of the British to the British poor both in Britain and in India. Legislation against public poverty in both India and Britain displayed an acute sense of class consciousness. The 1874 European Vagrancy Act passed in India was inspired by the same blend of charitable intent and disgust with the public exhibition of poverty as the British poor laws. The perception of vagrant leprosy sufferers as criminal in the Indian response to the 1889 Bill was entirely consistent with British perceptions of the European vagrant enshrined in the European Vagrancy Act. Under the European Vagrancy Act, the government workhouse became in effect a type of prison in which vagrants were subjected to punishment as if they were convicts. Penal sanctions were imposed on European vagrants for disobedience or leaving the workhouse without permission. Both the 1889 Leprosy Bill and the British and Indian discussions of its provisions showed signs of this tendency in nineteenth-century British thinking to criminalize the poor.

The British were more sympathetic to vagrancy caused by illness, which they distinguished from vagrancy due to indolence and vice.[34] The British attitude towards poverty cannot, however, be seen simply as a colonial imposition. The British link between poverty and loss of moral quality fitted squarely with Indian subjection of vagrant, unemployed leprosy sufferers to the full weight of caste opprobrium. The 1889 Leprosy Bill had little to do with racism and much to do with class, informed by notions of poverty as immoral. The Indian population was as severe with their own vagrant poor as the British were with theirs. It was with the sanction of Indian middle-class public opinion that the British sought to confine vagrant leprosy sufferers and ascribed to them the status of criminals.

Protection of religion and caste

Although the Government of India was prepared to allow class distinctions to become blurred in the 1889 Bill, in the more sensitive area of Indian religious observance the government was particularly careful to follow Indian public opinion. Religion was perceived as a more volatile area of Indian concern. Section 10 of the 1889 Bill declared: 'No leper shall against his will be sent under section 4 or section 5 to any retreat where attendance at any religious observance or at any instruction in religious subjects is obligatory on lepers accommodated therein.'

Moodeen Sheriff, honorary surgeon and foundation secretary of the Central Mahommedan Association, established in 1886, was stringent in his criticism. He pointed out that the Bill went against the Government of India's policy of strict religious neutrality and expressed surprise that the introduction of 'religious observance or instruction' could be countenanced by government 'in public retreats intended for the lepers of all classes and nations in this country'. He thus requested the Government of India not to 'allow any religious interference to exist in retreats for lepers' and suggested the omission of section 10 from the Bill.[35] Moodeen Sheriff's comments reflected the embattled position of the Muslim community in nineteenth-century south India. Lacking the political and economic power under British rule that in the past had made up for their low numbers, the Muslim community sought to maintain its identity through the formation of associations[36] and, as evident in Moodeen Sheriff's remarks, by protection of their religious integrity from state interference.

There was little comment from the Indian community regarding section 10 of the Bill. Only Sivangnana Muddalier warmly approved of section 10 'in so far as it implies a spirit of religious toleration'. However, linking the matter to protection of caste status, he recommended that the words 'or where his caste will be lost' should be added. Muttusami Aiyar was also troubled by the lack of caste recognition within the asylums. While having generally 'no objection to the proposed legislation', Muttusami Aiyar felt that: 'In order . . . to prevent irritation on the ground that caste prejudices are not respected in retreats, it is desirable to provide that facilities be afforded for lepers to conform to their caste rules and observances'.[37]

British authorities in Madras were sympathetic towards Indian concern to maintain caste distinctions within the asylum. As in the case of treatments for leprosy, in the Madras presidency, local officials were more sensitive than the central government to the immediate concerns and needs of leprosy patients. LeFanu, among others, endorsed the view of the 'native gentlemen' consulted in North Arcot district, that the law must make provision for 'recognizing caste' in the segregation of leprosy sufferers.[38]

The 1889 Bill was, in effect, more a measure to control vagrancy than to restrict the spread of leprosy. The provision for confinement of leprous vagrants in the Bill was devised to ensure that the destitute had some means of relief as well as to clear the Indian streets of the repulsive sight and nuisance of leprous beggars. Even so, the Bill's provision for forced confinement of vagrant leprosy sufferers completely contravened their desire to remain free. In Madras, the 1889 Bill was perceived as principally a vagrancy measure with little merit for leprosy control. However, the Bill did provide a context for the articulation of new class relationships developing within the Indian community and clarification of just how much intervention in Indian life the Madras government would sanction in a climate of nascent nationalism.

Madras government response

The Madras government ultimately rejected the 1889 Bill on the grounds that confining only vagrants with leprosy did not address satisfactorily the requirements of disease control. The Madras government emphasized that the 'danger of contagion . . . from those living in respectable houses, who marry and carry on business' was greater than from the 'wandering and outcaste' leprosy sufferers. The government concluded that only legislation requiring the compulsory removal of leprosy sufferers to asylums until they were either cured or dead would effectively control the disease, but that such a measure was impossible on financial grounds. As in the first part of the nineteenth century, financial considerations contributed to the Madras government's objection that the Bill enabled the well-to-do with leprosy to 'throw on the State the whole burden of the support for any period of time' without any guarantee that they would not return to the community and risk infecting others.

In agreement with all local governments and administrations except Hyderabad, the Madras government rejected the draft legislation.

They proposed, instead, an amendment to the Indian Penal Code preventing leprosy sufferers from working as butchers, bakers, shop keepers, barbers or washermen, occupations which were believed to spread the disease. The Madras government emphasized that their intention was management of leprosy, not containment of the sufferer. They stipulated that the exercise of such control be taken 'not with the object of generally prohibiting lepers from earning their own livelihood but with that of dealing with any special cases in which the danger of contagion is clearly shown to arise'.[39]

Indian government conclusions

In the light of the India-wide discussions of the Bill, the Government of India, comprising Lansdowne in council, revised its opinion that legislation to control leprosy was appropriate. Despite the fear of infection engendered by Fr Damien's death, the Government of India argued in its submission of the presidency responses to Viscount Cross, secretary of state for India, that information about the means of leprosy transmission was still too tenuous to justify legislation against all leprosy sufferers. Further, the Government of India noted that contrary to the understanding of most medical men, the Bill had not been framed on the assumption that leprosy was contagious. This partially explained the Indian government's willingness to protect the middle classes and focus on only the vagrant leprosy sufferers in the draft legislation. The Government of India cited Lewis and Cunningham's findings for the sanitary department in order to argue that there was no scientific justification for even a limited enactment against those with leprosy in 'an advanced and specially dangerous stage' or to encompass any 'dangerous lepers who might, after warning, persist in coming out and mixing with the public'. Reflecting the authority of the provincial administrations in the negotiation process, the Government of India bowed to pressure from the presidency governments to give disease control priority over social concerns in the formation of legislation against leprosy. The Government of India concluded that it was 'expedient to postpone legislation for the present' and decided to await the results of the Leprosy Commission to India for 'more definite information as to the causes of leprosy and the best means to be adopted for its prevention'.[40]

At each remove from the local Indian situation, government opinion reflected less the issue of containment of leprosy and more the concern to satisfy public opinion. Although the Government of India's letter of July 1890 to the secretary of state had emphasized the

Bill's inadequacy as a means of controlling the disease, Viscount Cross stressed the demands of public opinion in the presidency replies submitted by the Government of India. He argued strongly against the Indian government's grounds for postponing legislation. In opposition to the Government of India's conclusion that 'there is a very considerable divergence of opinion as to whether leprosy is contagious', he suggested that the pressure of 'public sentiment', evident in the provincial government's responses, was sufficient to warrant legislation for the segregation and confinement of leprosy sufferers. He also argued that delaying legislation until more definite information regarding the causes of leprosy became available was inappropriate since it had been already decided that legislation was necessary. He concluded that 'in view of the large body of Native opinion in favour of some legislation', and since the Leprosy Commission, though potentially 'interesting' could not guarantee 'any definite results', legislation of 'the very moderate character... under consideration' should be enacted.[41]

The Government of India, reflecting the authority of the provincial governments, responded critically to the secretary of state, reiterating the arguments against legislation put forward by the Madras and other provinces. The Government of India urged Viscount Cross to consider the requirements of public safety from the disease in preference to the support of Indian public opinion for legislation which, the Indian government asserted, was far weaker than he believed. Certainly there was not the immediate fearful reaction to Fr Damien's death in the Madras vernacular press that there had been in Britain. The Government of India endorsed the concern of the provincial governments for disease control, asserting that legislation for compulsory detention of leprosy sufferers would only be justifiable when it was established beyond reasonable doubt 'that the disease is contagious or at least very readily spread by accidental inoculation'. Further, the Government of India reaffirmed its commitment to the Leprosy Commission, expressing trust that the 'Commissioners... will at all events be able to throw some light upon this fundamental and much disputed point'.[42]

Even at the highest level of colonial authority, the development of legislation to control leprosy was subject to the processes of negotiation with the lower levels of government and the upper-middle-class Indian public. The poor and vagrant leprosy sufferers were the principal targets of the Government of India's legislation. However, the response to the 1889 Bill from the provincial governments widened the parameters of discussion to admit legislative measures against leprosy sufferers in employment who risked spreading the disease. By 1890, British and Indian opinion suggested that an Act tailored more specifically to disease control was more appropriate than what was effectively a Vagrancy Act. Provincial opinion, endorsed by the Government of India, prevailed against the secretary of state and legislation was postponed.

The Leprosy Commission's report

The report of the Leprosy Commission and of the committees which commented on its findings added to the cacophony of competing opinion on the confinement of leprosy sufferers in India. It is a reflection of the substantial change in British attitudes towards legislation against leprosy that the Royal College of Physicians' 1867 report had brought about the repeal of all existing legislation to confine leprosy sufferers in the British Empire, while the Leprosy Commission merely brought a postponement. The commissioners were requested to state in their final report their conclusions and supporting arguments concerning: 'the desirability or otherwise, first, of encouraging the voluntary partial withdrawal of lepers from amongst the non-leprous population; secondly of enforcing the complete isolation of all lepers; and, thirdly, of enforcing the isolation of certain lepers.' Further, they were instructed 'to describe minutely' what they believed to be the best plans for ensuring the efficient practical implementation of their recommendations. The Government of India, however, reserved the right to accept or reject their recommendations.[43]

The Leprosy Commission's report represented the late-nineteenth-century understandings in India of the aetiology and pathology of leprosy just as the Royal College of Physicians' enquiry into leprosy had embodied mid-nineteenth century conceptualizations of the disease. On the basis of its enquiries, the Leprosy Commission decided that: 'Leprosy is a disease *sui generis*,' quite distinct from syphilis or tuberculosis and 'is not diffused by hereditary transmission'. The commission found that leprosy was 'inoculable', did not originate directly from any particular kind of food, specific climate and telluric conditions or insanitary surroundings, and neither did it affect any race or caste in particular. However, insanitary surroundings, poverty, poor ventilation and drainage, 'by causing a predisposition increase the susceptibility of the individual to the disease'.

In most cases, the commission decided that leprosy: 'originates *de novo*, that is, from a sequence or concurrence of causes and conditions, . . . which are related to each other in ways at present imperfectly known'. The commission reported that leprosy was not easily or efficiently transmitted. The extent to which leprosy was propagated by contagion and inoculation was found to be 'exceedingly small' and the lack of hereditary transmission combined with 'the established amount of sterility among lepers' meant that 'the disease has a natural tendency to die out'.[44]

Since their findings indicated that leprosy was not a highly contagious disease, the commissioners proposed three practical measures for the control or restriction of leprosy in India. They opposed an Imperial Act specifically directed against leprosy sufferers since, in their view, 'neither compulsory nor voluntary segregation would at present effectually stamp out the disease, or even markedly diminish the leper population, under the existing conditions of life in India'. Rather, they suggested that 'lepers and leprosy' be regulated by means of bye-laws. Consistent with the tenor of debates over the 1889 Bill, the commission made a distinction in its recommendations between the control of the disease of leprosy and that of vagrants with leprosy, proposing a measure to prevent the spread of leprosy through employment distinct from that to prevent begging by leprous vagrants. The commissioners recommended that leprosy sufferers be prohibited from 'the sale of articles of food and drink . . . practising prostitution and . . . following such occupations as those of barber and washerman, which concern the food, drink and clothing of the people generally'. Further, the commissioners advised that municipal bye-laws be passed, 'preventing vagrants suffering from leprosy from begging in or frequenting places of public resort, or using public conveyances', thereby discouraging leprosy sufferers from congregating in towns and cities. To accommodate leprosy sufferers, the commissioners suggested that existing asylums be enlarged by municipal or private funds and new asylums be built near towns which had no such facilities. After consultation with medical authorities, police would be empowered to order those breaching regulations either to return to their homes or to enter an asylum.[45]

W.R. Rice, surgeon-general for India, concurred with the commission's findings on the contagiousness of leprosy. However, the executive committee of the National Leprosy Fund and the special committee appointed specifically to consider the commissions' report,[46] despite the wealth of evidence assembled by the Leprosy Commission, rejected their findings that 'the extent to which leprosy is propagated by contagion and inoculation is exceedingly small' and that leprosy 'in the majority of cases originates *de novo*'. In opposition to the Leprosy Commission, the two committees supported a Government of India Act to control leprosy sufferers. While generally supporting the commission's recommendations for 'the regulation of lepers and leprosy in India', the special committee did not agree that municipal bye-laws 'will be necessarily or universally the best means of effecting that object'. Instead, reflecting Carter's continued influence in medical circles, they endorsed his arguments for the planned establishment of asylums to serve several districts as refuges and places of detention, and for using legislative authority to ensure the strict isolation of leprosy sufferers in the home or their removal and isolation from public areas. In short, the two committees wished for a greater level of centralized control over the actions of leprosy sufferers than that advocated by the Leprosy Commission, and strongly encouraged the Government of India to take legislative action for the segregation of leprosy sufferers rather than leaving such steps in the hands of the municipalities.[47]

The members of the two committees were not, however, entirely in accord. The five members of the National Leprosy Fund executive committee who had medical credentials, Sir Andrew Clark, Sir W. Guyer Hunter, Sir James Paget, Sir Joseph Fayrer and Jonathan Hutchinson, dissented from the two committees and supported the Leprosy Commission's recommendations.[48] It was this view which ultimately gained Government of India endorsement. The majority of the local governments and administrations also endorsed the Leprosy Commission's recommendations, giving further grounds for Government of India support. Noting with respect the opinion of the special committee and the executive committee of the National Leprosy Fund, the Government of India nevertheless felt 'constrained to accept' the commissioners' conclusions, since they were supported by the 'distinguished medical authorities', cited above, and 'by the general opinion of those in India who are qualified to pass judgement on it'. The Government of India agreed that there was insufficient evidence of the contagiousness of leprosy to justify an Imperial Act, and that the management of leprosy should remain in the hands of the municipalities.[49]

The local governments and administrations unanimously supported the Leprosy Commission's first recommendation to prevent leprosy sufferers following specific trades and, with the exception of the Madras government and the chief commissioner of Assam, generally approved of 'sending to, and retaining in, asylums vagrant pauper lepers who congregate in towns'. The Madras government, consistent with their attention to disease control as a concern separate from vagrancy control, concluded that the Leprosy Commission's findings were sufficient grounds for 'refusal to assent to any measures of compulsory segregation or of isolation and supervision of lepers in their homes'. The Madras government supported the use of municipal bye-laws to prohibit leprosy sufferers from following specific trades as suggested by the commissioners. However, since no measures existed to prevent the transmission of syphilis through prostitution, the Madras government considered inappropriate the Leprosy Commission's proposed regulation of prostitution to control leprosy, from which there was far less threat of contagion. The Madras government approved limiting the concentration of vagrants suffering from leprosy in towns and public places and on public transport. However, rather than the forced confinement of vagrant leprosy sufferers as recommended by the Leprosy Commission, the Madras government proposed using municipal bye-laws only to expel leprosy sufferers from towns and cities. Further, the Madras government allowed use of local or municipal funds for the establishment and maintenance of asylums, as suggested by the commissioners. However, given 'the very infinitesimal danger of contagion and inoculation and the apparent decrease of the disease' they considered funding asylums a low priority. Rather than directing funds for general medical relief to building leprosy asylums, the Madras government preferred to leave provision of asylums mainly to private charity. They insisted, with the commission, that in every instance of legal interference with leprosy sufferers, 'competent medical authority' be consulted before taking action.[50]

The Government of India, reviewing the Leprosy Commission's report and the responses of the provincial governments, ultimately agreed with the commission. They expressed, in their resolution of 23 March 1895, their sense of assurance 'that the extent to which [leprosy] is propagated in India by contagion is small' and resolved that they were 'unable to approve of the compulsory segregation, either absolute or partial, of lepers except under special circumstances', that is, where asylums or farms which provided for the hygienic and health needs of leprosy sufferers were available. The Government of India followed the commissioner's recommendation that a leprosy sufferer found begging in town should be either returned home or confined in an asylum and that the power to confine vagrant leprosy sufferers should rest with the municipalities. Further, they approved the application of municipal bye-laws to prevent leprosy sufferers participating in particular employments. The Government of India distinguished between the medical requirements of regulating leprosy and the requirements of public opinion. They made it clear that, strictly on the basis of the very minimal threat of contagion from leprosy sufferers, regulation was unjustified but that in view of the public fear of encountering leprosy, and the 'danger of septic poisoning from any . . . running sore or ulcer' such regulation through municipal bye-laws was necessary.[51]

By the close of 1895, the status of legislation to control leprosy had been renegotiated. Two subjects of control were clearly identified: leprosy sufferers and leprosy. The Madras government had become stronger in its resistance to forcible confinement, while the Government of India, approving of the practice under limited circumstances, had delegated responsibility for its legislative implementation to the local governments and administrations. The Government of India was prepared to admit a multiplicity of responses to leprosy and to limit legislative authority over leprosy sufferers and leprosy to the local municipal level. Whether the 'leper retreat' was to be a hospital or a prison was left to the local governments and administrations to decide.

The 1896 Leprosy Bill

The Government of India's resolve to avoid an Imperial Act to control leprosy did not hold for long. In 1895 an act was passed in Bengal for the segregation of paupers with leprosy, preventing leprosy sufferers pursuing particular employments and facilitating funding for maintenance of leprosy asylums. The Bengal Act included specific restrictions. It could only be introduced in an area in which there was already a leprosy asylum, only beggars suffering from leprosy who had no other means of subsistence could be confined, removal to an asylum was conditional on a certificate from a medical authority and the leprosy sufferer could appeal against the order for his or her confinement. Further, the forms of employment from which the leprosy sufferer was excluded were expressly stated. Given the interest in Burma and other provinces in applying the Bengal Act, the Government of India proposed that a general Act be devised which could be applied in the provinces at the discretion of each local government. Accordingly, the 1896 Leprosy Bill, modelled on the Bengal Act, was introduced into the Legislative Council on 30 July 1896 and by August 1896 was under consideration by the provincial governments. The 1896 Bill was put forward on the grounds that the Leprosy Commission had argued for the expediency of segregating pauper leprosy sufferers and the prohibition of leprosy sufferers from pursuing specific employments. In the 'Statement of Objects and Reasons' for the Bill, there was no mention of the Leprosy Commission's conclusion that there was insufficient evidence of the contagiousness of leprosy to justify legislation segregating leprosy sufferers.[52]

In the Madras presidency, as J.N. Atkinson, collector of Kistna, commented, it was still unclear 'whether the ultimate object of the Bill is to protect society from risk of the spread of a loathsome and incurable disease, or from the mere annoyance and disgust entailed by the public exhibition thereof'. While fully aware of the discrepancy between the Leprosy Commission's recommendations and the Government of India's Bill, Atkinson argued that the Bill was justifiable 'either on the hypothesis that leprosy is a contagious disease, or on the ground of the admittedly horrible nature of the scourge'.[53] Although the Government of India had resolved in 1895 that fear of contagion was unwarranted, by 1896 the offensiveness of leprosy to public feeling was given equivalent status to medical concerns in the Madras presidency's support for legislation at Government of India level. This was a significant departure from the debate surrounding the 1889 Bill and the Madras government's response to the Leprosy Commission's Report. At that time, there had been considerable tension between the social and medical aspects of leprosy, and the Madras government had resisted legislation for confinement of leprosy sufferers on social grounds.

Despite their previous opposition to an Imperial Act to control leprosy, significant differences from the 1889 Bill made the 1896 Leprosy Bill more acceptable to the Madras government. The 1896 Bill generally met with approval from the provincial governments, with criticism confined to Section 8 concerning the confinement of paupers with leprosy and sections 9 and 10 regarding the prohibition of leprosy sufferers from engaging in certain employments, using public water supplies and travelling on public transport.[54]

Specific inclusion of measures to regulate leprosy sufferers in employment met a principal criticism of the 1889 Bill. While, as with that Bill, confinement was advocated primarily for the pauper leprosy sufferer, by 1896 the Government of India had decided that limitations on the 'liberty' not only of vagrant leprosy sufferers but of those in employment was in the public interest. Madras medical authorities were in agreement. As Lieutenant-Colonel W.A. Lee, surgeon, noted:

> It cannot be held to be an undue interference with the liberty of the individual to check the unrestrained freedom now enjoyed by lepers in the pursuit of avocations conducive to the dessemination of their malady, or its transfer to the healthy by direct contagion'.[55]

Even so, confinement of leprosy sufferers in employment was advocated only as a last resort, subsequent to a fine and a warning for a first offence.[56] Although the principle of restriction was established, there was considerable debate over which specific occupations should be prohibited to leprosy sufferers. The Madras government recommended that teacher, cigar maker, compounder of medicine, butcher, dairyman and vaccinator be added to the list of activities excluded under the Bill.[57]

The 1896 Bill allowed fuller intervention in the daily life of leprosy sufferers than its predecessor, both to prevent public exposure to the repulsiveness of the disease and to reduce the risk of infection. In addition to regulating leprosy sufferers in employment, a leprosy sufferer who 'bathes at a public well or bathing-place or takes water from any public drinking fountain, well, hydrant, tank or reservoir' or who 'rides in any public conveyance', was to be fined Rs 50.[58] These aspects of the 1896 Bill attracted considerable criticism from both the British and the Indian officials consulted. In particular, the restrictions placed on access to water for leprosy sufferers were considered to be too stringent. As G.W. Elphinstone, district magistrate for Godavari, commented, by the provisions of the Act, 'it is practically made a criminal offence for a leper to drink water'. He continued that in most large towns the only sources of drinking water are usually public and 'circumstances can easily be imagined under which a leper could not get anyone to draw water for him'.[59] Even so, the Madras government endorsed these restrictions on access to water and recommended their extension to include public wells and bathing places open to public use.

The limitations on the use of public transport by leprosy sufferers were also considered to be too severe. The Government of India's Railways Act, 1879, devised to prevent the spread of all 'infectious or contagious disorders', and superseded by the 1890 Act, had already provided legislation to regulate rail travel by leprosy sufferers. In addition, the Madras government had its own regulations requiring those suffering from leprosy of an offensive type to travel in a compartment separate from the other passengers. In practice, however, the regulations made little difference to the movement of leprosy sufferers who frequently travelled on the South Indian Railway 'without being compelled to take even a compartment.'[60]

Muhammad Raza Khan, joint magistrate for South Arcot, suggested that if a leprosy sufferer were to be penalized for riding in a public conveyance, 'some ameliorating arrangements for him to perform his journey' should be made. The Madras government agreed that limitations on use of transport by leprosy sufferers were excessive, noting that the practical danger of contagion from such exposure to leprosy was minimal. The Madras government proposed that in the case of railway carriages at least, the restrictions be removed and the leprosy sufferer only be required 'to give notice to the railway authorities so as to enable them to take steps to have the carriage swept or washed out with some disinfectants'.[61]

The 1896 Bill raised the same complex of issues as the 1889 Bill. Pauper leprosy sufferers appeared to have less claim to liberty than those of higher socio-economic status and there was tension over whether the leprosy asylum was to be a hospital or a prison. As in the earlier debates, some Madras medical authorities recognized that pauper leprosy sufferers had a degree of autonomy and even a valid socio-economic status. W.B. Browning, surgeon to the governor, noted that the arrested and incarcerated beggar with leprosy was often 'the only bread-winner of a family' rather than a burden and a public nuisance, and questioned the lack of provision made for the sufferer's family in the Bill. Even so, Browning did not go so far as to argue that leprosy sufferers who were poor had rights to liberty equivalent to those of a higher socio-economic status. Rather, he continued, restriction of liberty was appropriate on charitable grounds: 'as one is dealing generally with poor, uneducated people of a low order of intelligence, that it is justifiable, in their own interests chiefly, to use a gentle compulsion in isolating and treating them'. Browning was at the same time concerned that with legislation to control leprosy, the Leprosy Commission's benevolent intention to exclude leprosy sufferers from towns and place them in country asylums with the benefits of hygiene and good food would be lost. He continued that 'leper asylums might come to be regarded as prisons instead of being looked upon as havens of refuge by those suffering from the disease'.[62]

As during the previous debates, the Madras Indian upper-middle class argued for greater protection of leprosy sufferers from abuses of the power to confine. C. Jambulingam Mudaliyar, a pleader in Cuddalore, argued that provision should be made in the 1896 Bill, as in the Criminal Procedure Code, that the leprosy sufferer should be brought before a magistrate within 24 hours of arrest 'to prevent unnecessary delay and harassment to people'. Under Section 12 of the 1896 Bill, inspection was required for recent admissions 'as far as circumstances will permit', in effect making it optional to inspect leprosy sufferers subsequent to their admission. Jambulingam argued that the prompt inspection of each leprosy sufferer, once confined in the asylum, should be compulsory to prevent mistaken arrest. The comments by C. Jambulingam Mudaliar, who was the contact between the Madras Mahajana Sabha and the local political organization in Cuddalore, reflected the continued interest not only of the Madras metropolitan professional classes but of the *mofussil* professionals in determining the degree of British interference in Indian social life acceptable to the Indian upper-middle classes.

While Jambulingam's comments were concerned as much with the protection from wrongful confinement of those who were not paupers as of those who were not leprosy sufferers, the Indian upper-middle class also expressed the need for protection of pauper leprosy sufferers from unjust confinement. Many Indian gentlemen considered that withholding from the pauper leprosy sufferer the right of appeal against confinement under Section 14 of the act, was 'unnecessarily severe on an unfortunate class'. According to Section 14, any other person declared to be a leprosy sufferer by an 'inspector of lepers' was allowed the right of appeal against the finding.

The permanence of confinement imposed under the Bill was also a point of dispute for the Indian gentlemen. N. Subbarao Pantulu objected that under the act, once a pauper with leprosy was committed to an asylum, 'there is no express provision for his release in cases where he subsequently ceases to be a pauper or where his relatives subsequently came forward and guarantee to support to him'. Subbarao Pantulu proposed that the inclusion of provisions for release of pauper leprosy sufferers in such circumstances be included in the Bill. Further, he regarded as excessive the punishments for those leprosy sufferers who twice breached the provisions of Section 9 regarding engagement in employment, use of public water and of public transport. Under Section 10 of the Bill, leprosy sufferers could be banished from their home or confined in an asylum for their second offence. Subbarao Pantulu proposed that both banishment and detention should be limited by the magistrate to a fixed term, no greater than a year.[63] In both matters addressed by Subbarao Pantulu, the same concern to protect distinctions between the wider middle class and the vagrant poor was evident as in discussions on the 1889 Bill. Subbarao Pantulu recommended release of pauper leprosy sufferers when their socio-economic circumstances had improved, not when they were freed of their disease. Similarly, he considered that those confined for breaching restrictions on the activities of leprosy sufferers, rather than for pauperism, should not be subject to the same lifetime confinement.

To some extent the Madras government incorporated protection of pauper leprosy sufferers into their response to the 1896 Bill. As in the previous debates, the Madras government sought to ensure that the power of arrest was exercised with a significant level of responsibility and thus recommended that only an officer in charge of a police station should have the power to arrest paupers believed to have leprosy. The Madras government followed C. Jambulingam Mudaliyar's recommendations, inserting a clause to protect those arrested on suspicion of leprosy from unnecessary detention after arrest and ensuring swift clarification as to whether they were to be committed to an asylum or released. Consistent with Indian protection of class interests, the protective measures recommended by the Madras government emphasized that the arrest of pauper leprosy sufferers was more on account of their pauperism than their disease. The Madras government recommended that a person detained as a 'pauper leper' be discharged when his or her socio-economic status had altered as in the circumstances envisaged by Subbarao Pantulu.

In contrast with Indian upper-middle-class opinion, the Madras government did not approve of extending the right of appeal against confinement to pauper leprosy sufferers, reserving it for those confined for breach of regulations to prevent contagion. In the view of both the British and Indian upper-middle class, liberty thus continued to be class-specific. The Madras government was, however, generally tougher than the Indian gentlemen in areas where there appeared to be some threat of contagion. For those who violated the restrictions on employment, travel on public transport and use of water, the Madras government generally approved, and in some cases suggested, tightening provisions. Further, they did not approve of the measure, proposed by Subbarao Pantulu, to limit to one year the confinement of those who breached these regulations a second time.[64]

The 1898 Lepers Act

The Lepers Act was passed in 1898 and incorporated the substantial amendments recommended by the Madras government. To accommodate the variety of opinion from the local governments and administrations on the measures restraining employment of leprosy sufferers and their use of public transport and water facilities, the 1898 Act allowed local governments discretion in applying the provisions listed in Section 9 of the 1896 Bill. Consistent with the 1896 Bill, there was no reference to hospitalization of voluntary leprosy sufferers under the 1898 Act, reflecting the concern expressed by both Indian and British authorities to keep clear the distinction between patient and prisoner. Those who had been compulsorily confined under the Act and who left the leprosy asylum without permission were subject to re-arrest and return.[65] Under the Lepers Act, the leprosy hospital was, in effect, a prison for vagrants suffering from leprosy, similar to the workhouse under the European Vagrancy Act.

The belief that segregation would reduce the spread of the disease, a belief endorsed by the First International Leprosy Conference of 1897, prompted other countries to pass legislation for the confinement of leprosy sufferers. In addition to the Norwegian legislation of 1877 and 1885, compulsory segregation laws were enacted in 1883 in New South Wales, Australia. In Cape Colony, the Leprosy Repression Act was passed in 1885 after the South African Leprosy Commission reported an increase in leprosy through all areas of the population. However, the law was not promulgated until 1892, after further enquiry in 1889 revealed that the disease was continuing to increase.[66] India was not the only colony to hesitate in applying legislation to contain leprosy.

Generally, the Lepers Act and its predecessors were consistent with Indian attitudes to leprosy sufferers. The Lepers Act accorded with traditional Hindu Brahmin law concerning leprosy. In the first part of the Act, a 'leper' was defined as 'any person suffering from any variety of leprosy in whom the process of ulceration has commenced'. As S. Swaminatha Aiyar mentioned in his comments on the 1896 Bill, it was only those with the 'virulent and aggravated type' of leprosy believed to be incurable and usually identified as ulcerous leprosy, who, under the *dharmaśāstra*, were denied inheritance.[67] In targeting paupers with leprosy for forced confinement, the Bill was clearly in agreement with Indian class feeling.

Application of the Act in the Madras presidency

Although the 1896 Bill had generally met with approval and the Madras government's recommended amendments were largely incorporated in the 1898 Act, it was not applied in south India during the nineteenth century. The Act was immediately applied in Burma, except the Shan states, and in the districts of Allahabad, Benares, Lucknow and the Kumaon Divisions of the North Western Provinces, with Port Blair in the Andaman Islands following in 1900. In 1901 in Bengal, it replaced Act V of 1895. The Act was applied in Assam in the same year, in specific districts of the Central Provinces in 1903 and in the Dehra Dun and Cawnpur districts of the United Provinces in 1906 and 1909 respectively. Extension of the Act was proposed for the city of Madras only in 1910, and after protracted discussion, it was agreed in 1913 that the Act be applied to the entire presidency,[68] a decision which, though reported in the *Lancet* on the 10th of May, attracted remarkably little response in the vernacular and English Madras press.

Prevention of leprosy sufferers from pursuing specific employments was largely managed through municipal bye-laws framed under Section 409 of the Madras City Municipal Act, 1904. From 1908 to 1911, five leprosy sufferers were prosecuted under municipal bye-laws and some were fined between Rs 1 and Rs 5 for selling articles of food. The existence of comprehensive bye-laws and the low number of pauper leprosy sufferers at large in Madras City – the commissioner of police reported 16 and the superintendent of census operations only 12 – no doubt contributed to the lack of urgency in applying the Lepers Act. A further source of delay was the lack of suitable asylum accommodation. The decision to extend the Lepers Act throughout the presidency was dependent on the establishment of an asylum outside the city for segregation of arrested pauper leprosy sufferers.[69] By 1921 however, the Lepers Act was still not in force. Rather, in 1939 the Madras Public Health Act was promulgated including provisions preventing a leprosy sufferer appearing in a public place, travelling by public transport or having any form of direct or incidental contact with a member of the public. The Public Health Act was amended in 1944, increasing the restrictions on the public activities of leprosy sufferers. Most significantly, it provided that, where adequate accommodation was available, a specific area could be declared a 'segregation area'. All those with leprosy within the area could either voluntarily segregate themselves in the accommodation provided or would be forcibly confined. As under the Lepers Act, those who left confinement without permission were to be arrested and returned. The critical difference between the Madras Public Health Act, 1944 and the Lepers Act was its inclusion of leprosy sufferers of every class in its provisions for compulsory confinement, in the interests of disease control. However, like the Lepers Act, it was impossible to enforce.[70] In 1957 the Lepers Act was finally extended to the transferred territories in the state of Tamil Nadu by T.N. Act XXII, just subsequent to the initiation of the National Leprosy Control Scheme in 1955–66.

Effects of the Lepers Act

Ironically, after the enormous expenditure of energy in the discussion of legislation to confine leprosy sufferers, the Act, where it was applied, had little impact on leprosy sufferers in India. Where there was an increase in the number of pauper leprosy sufferers confined, the increase was largely due to voluntary confinement rather than police action. Many, fearing that they would be confined for life, preferred to enter asylums as voluntary in-patients and then depart once their ulceration appeared healed. This fear among pauper leprosy sufferers was, however, largely unfounded. Where the Act had been introduced it was only leniently applied. Enforcing the Act was not a police priority.[71] The officials concerned were overworked and had little time to attend to leprosy sufferers either as people requiring charity or as a public nuisance.

Further, in some areas the Act was interpreted to mean that pauper leprosy sufferers could only be confined while they had open sores. Once the ulceration had healed they were free to leave. Police thus felt their efforts were futile and had little incentive to continue their arrests. This inadequacy in the definition of 'leper' under the law had been pointed out by the Madras government as early as discussions of the 1896 Bill. From 1909, the Mission to Lepers similarly argued that the definition of 'leper' under the Act should be updated.[72] New medical evidence indicated that the majority of leprosy sufferers discharged bacteria through the nasal mucosa, irrespective of whether or not they had cutaneous lesions. The Lepers Act was amended in March 1920, redefining 'leper' as 'any person suffering from any variety of leprosy'; a wider definition of leprosy was also included in the Madras Public Health Act.[73]

The precise provisions of the Lepers Act, and indeed any law for the regulation or confinement of leprosy sufferers, were, however, largely irrelevant when the police disliked both apprehending and handling leprosy sufferers. As A. Donald Miller, secretary for the Mission to Lepers in India, observed four years later, it was only after the Act was applied that the difficulties in its implementation became evident: 'The human factor began to be recognized; it was seen that the normal policeman is not keen to arrest beggar lepers; it was seen that the beggar leper is not keen to be arrested and interned; and above all, costs loomed ominous.'[74]

The principal consequence of the Lepers Act related not to the requirements of disease control but to the costs of leprosy care. To help defray the costs of confinement under the Lepers Act, local and presidency governments developed a closer relationship with the Mission to Lepers. Under Section III of the Act, the local government could declare any place to be a 'leper asylum' for the purposes of the Act. In complying with the Act, presidency governments intended to leave the segregation and care of leprosy sufferers to private institutions as far as possible, supporting them with grants in aid. Since most private institutions for leprosy care existing in India by 1898 were run under the auspices of the Mission to Lepers, the passing of the Lepers Act began the process of delineating the new relationship between the Mission and government. The Mission insisted on retaining autonomy while receiving government funding linked to their provision of accommodation under the Act.[75] Thus, from the early twentieth century, care and treatment of leprosy sufferers substantially became the responsibility of the Mission to Lepers.

From being almost entirely a government matter, leprosy care became a missionary concern, a rare transfer of responsibility for indigenous leprosy sufferers from 'local state authority to western religious private philanthropy'.[76] The transfer was all the more remarkable for occurring at a time when missionaries were increasingly perceived in India as an extension of empire and as supporters of colonial rule against the forces of nationalism.[77] That such a shift in responsibility occurred reflected the economic burden of leprosy care on the state and, no doubt, also a British conviction that there would be no objection from the Indian upper-middle class to the involvement of missionaries with leprosy sufferers, particularly the poor with leprosy. Under missionary care, the leprosy sufferers' spiritual welfare, understood in Christian terms, became at least as important as their physical and emotional care. As Miller wrote in his history of the mission's early years: 'The total man called for total help in his total need'.[78] With the Mission to Lepers, a new phase in the history of leprosy in India begins and this history ends.

The Government of India's decision to confine leprosy sufferers was not the expression of a unified stance on confinement but was negotiated between different levels of medical, legal and government authorities and Indian upper-middle-class opinion. Disagreement and negotiation were essential to the construction of the Act and thus, the formation of colonial power over leprosy sufferers, incomplete and ineffectual as it was. Few presidency governments, the Madras government included, endorsed the Act they had helped to shape. Ultimately, it was the will of the police and the co-operation of the leprosy sufferers which were critical to the law's success. In their absence, confinement could not occur. The history of confinement in south India confirms Ignatieff's criticisms of Foucault's analysis of power. It is not just those in authority who formulate power but those responsible for its local application and the subjects of power who are a 'constitutive element in the history of power'.[79] Despite the rhetoric of incarceration, legislation against leprosy was ineffective both as disease control and as vagrancy control. Resistance by legal officers and leprosy sufferers to its application, combined with the costs of funding asylums, made the Lepers Act a dead letter. Apart from satisfying the moral demands of imperial care for British subjects and providing a forum in which the Indian upper-middle class could express their solidarity with the values of the educated rulers, the Lepers Act did little for the colonial cause. In the context of leprosy control in nineteenth-century south India, law was as ineffectual a 'tool of empire' as medicine.

Conclusion

The history of leprosy in south India not only provides a fascinating insight into an endemic disease. It offers a context for the detailed examination of the political, economic and social life of south India in the nineteenth century and the exploration of global issues such as the stigmatism of disease, the development of scientific medicine, the articulation of class distinctions and the nature and functioning of colonial power.

The common beliefs, both refuted by Gussow, that leprosy was stigmatized equally in all cultures and that modern leprosy stigma is part of a continuous tradition, are further questioned by the indigenous and British response to leprosy in south India. In indigenous culture there was a disjunction between the degree of stigma associated with leprosy in the textual religious and legal tradition, particularly the vulnerability of the leprosy sufferer to loss of inheritance and outcasting, and the practice of ostracism which targeted only the poor with leprosy. Many Indian leprosy sufferers continued to live and work among family and community without hindrance. Municipal legislation in Madras, intended to restrict the activities of leprosy sufferers in specific areas of employment, was rarely used. Similarly, both the Madras Public Health Act of 1939 and the Lepers Act of 1898 made little actual impact on the working lives of leprosy sufferers. Within the textual tradition itself there was also disagreement over the precise degree of ostracism appropriate to leprosy sufferers. Further, while leprosy sufferers were often abhorred by the British in south India, many British medical officers perceived leprosy in medical rather than moral terms and emphasized that the old stereotype of the leprosy sufferer as dissolute was not born out by observation.

The exploration of fundamental questions concerning the nature of leprosy and the response to leprosy sufferers in nineteenth-century south India thus contributes to the understanding of the history of leprosy stigma in global terms. Similarly, discussion of the medical and legal aspects of leprosy in south India provides insight into the broader relationship between colonized and colonizer and the nature of power as it functions in the imperial context.

Arnold and others have argued convincingly that from the 1830s, the British in India identified their medicine as distinct from and superior to indigenous medicine. Consequently, state-sanctioned medicine increasingly neglected indigenous medicine, and the emerging British 'scientific' medicine came to dominate medical practice in India. The history of leprosy treatment in south India, however, suggests some qualification of this view. Despite the prominence of European- and British-developed cures from the 1870s, the dominance of British medicine was by no means complete. In leprosy treatment, where cure proved elusive, the British remained willing to borrow from Indian medicine throughout the nineteenth century. Further, by the early twentieth century, the indigenous remedy chaulmugra oil, rather than the European- and British-developed treatments, had become the basis of leprosy treatment in India and internationally.

The development of a scientific medical culture in India occurred in the context of this dynamic relationship between indigenous and British medicine. In the medical treatment of leprosy, ultimately the opinion of local medical officers prevailed over the treatments endorsed by the British and Indian governments. Similarly, while Europe prompted scientific enquiry into leprosy in India and the first large-scale colonial investigation of leprosy was devised by the Royal College of Physicians, London, leprosy research in India also contributed to the development of a colonial medical culture which was related to, but distinct from, European and British medicine. The history of leprosy research in India, particularly the influence of local and presidency medical authorities on the trials of leprosy treatments, supports Arnold and Harrison's view that British medicine in India was neither entirely derivative of medical science at the centre of empire nor dependent on it for direction. British leprosy treatment, was, throughout the nineteenth century, stimulated and at times circumscribed by presidency and local medical initiatives. Medical science at the 'periphery', guided by local knowledge, which included, to some extent, indigenous medical knowledge, had its own priorities.

British medicine was too under-funded and too haphazard in its administration ever to be an effective 'tool of empire'. Opportunities for greater levels of medical research and intervention, often requested by medical officers in direct contact with the Indian sick, were not taken up at presidency and higher levels of government because Indian health was not an economic priority. British medicine was not an inexorable source of colonial power, radiating from the 'centre' to the 'periphery'. Rather, at each remove from Britain, at the Government of India, presidency and local levels of medical authority, the capacity of British medicine to act as an expression of colonial power became more diffuse, and more subject to the influence of British medical authority in India – and thus to negotiation and opposition from within British medical ranks. At the local level of direct contact between doctor and patient, British medicine was ultimately subordinate to the wishes of the leprosy patients, the majority of whom were Indian.

A similar diffusion of power is evident in the British legal response to leprosy, particularly in the use of law to enforce confinement of leprosy sufferers. Decision-making was subject to fissures in the structures and ideologies of the state. Competition and conflict inhibiting unity between religious, medical and legal authorities, differing shades of public opinion and economic imperatives all influenced who was to be confined in south India and on what terms. In addition to conflict and resistance among authorities, resistance by those targeted for confinement was a critical 'constitutive element in the history of power'. The effectiveness of the decisions made about those with leprosy was entirely dependent upon the co-operation of the sufferer. Law is as limited a 'tool of empire' as medicine. British power in India was shaped by resistance and negotiation, often for political advantage, conducted on medical, moral, religious and economic grounds.

The issues raised in this book – the notion of an excluded population for whom confinement is seen as appropriate; the exercise of power against these peoples by those who place themselves at the 'centre' rather than the 'periphery' of the community; the fragmentation and shifting allegiances inherent in power despite its appearance of unity; and the autonomy of those living on the margins and their capacity to resist and negotiate freedoms – these issues are not just relevant to nineteenth-century British colonial Madras. They are critical to the social history of any community.

Most leprosy sufferers of south India escaped confinement. However, confined or free, the medical, legal and cultural structures of their own communities and of British India impacted profoundly on their lives. More than a part of the history of Madras, this volume emphasizes the essential humanity of the leprosy sufferer, despite the disfigurement of disease, and confirms the experience of suffering as intrinsic to human history.

Biographies

Whitelaw Ainslie[1]

Whitelaw Ainslie (1768–1837) entered the Madras Medical Service as assistant surgeon in 1798, was promoted to surgeon in 1794 and superintending surgeon in 1823, retiring in 1828. He was well regarded for his many writings on medical subjects, particularly the *Materia Indica*, and was knighted for professional eminence in June 1835. Ainslie was author of *The Use of Balsam of Peru*, 1811; *Edible Vegetables*, 1811; *Materia Medica of Hindustan*, 1813, which was enlarged and republished as *Materia Indica* in 1826; *Observations on Cholera*, 1825 and *Medical Observations* (in Murray's British India), 1832. In addition, Ainslie published two non-medical works, a drama entitled *Clemanza, or the Tuscan Orphan*, in 1822 and *Historical Sketch of Introduction of Christianity into India*, 1835.

Claude Bernard[2]

Claude Bernard (1813–78) was renowned for his skill in dissection, demonstration, experimentation and, more controversially, vivisection of animals. He was a founder of the Société de Biologie in 1848 and was elected perpetual president in 1867. Bernard received the title Chevalier of the Legion of Honor in 1849, and in 1855 a chair of general philosophy was created for him at the Sorbonne. In 1864 he established a laboratory in which to pursue his researches in physiology. Bernard made considerable advances in physiological understanding of the functions of the pancreas and liver and of the sensory and motor properties of nerves.

Henry Vandyke Carter[3]

Henry Vandyke Carter (1831–97) became a Member of the Royal College of Surgeons and received his Licentiate of the Society of Apothecaries in 1852. In 1856 he obtained his MD from London University. He joined the Bombay Medical Service as assistant surgeon in 1858 and was promoted to surgeon in 1870. He advanced to the post of surgeon-major in 1873 and brigade surgeon in 1882 before retiring in 1888. He was appointed honorary deputy surgeon-general and the Queen's honorary surgeon on 8 November 1890.

William Robert Cornish[4]

William Robert Cornish (1828–97) became a Member of the Royal College of Surgeons in 1852 and joined the Madras Medical Service in 1854. He was promoted to surgeon in 1866 and became a Fellow of the Royal College of Surgeons in 1868. Cornish was sanitary commissioner for Madras from 1870 to 1875 and again from 1876 to 1879. He was promoted to surgeon-major in 1873 and was appointed brigade surgeon in 1879. In recognition of his ability he was promoted to the rank of surgeon-general in 1880 over the four deputy surgeon-generals and seven senior brigade surgeons. Cornish was made a Companion of the Order of the Indian Empire in 1880, and from 1885 received the Good Service Pension. He was author of *Prison Dietary and Food* published in 1863; *Typhoid Fever in Madras Presidency*; *The Cleansing of Indian Towns*; *Madras Medical and Sanitary Regulations* of 1870 and *Cholera in Southern India*, 1871.

James MacNabb Cuningham[5]

James MacNabb Cuningham (1829–1905) gained his MD from Edinburgh in 1851 and entered the Bengal Medical Service as assistant surgeon in the same year. He was promoted to surgeon in 1864, surgeon major in 1871 and surgeon general in 1880, retiring in 1885. Cuningham was sanitary commissioner with the Government of India from 1868–84. Cuningham was made Companion of the Order of the Star of India in 1885 and Queen's Honorary Surgeon in 1888. He obtained the degree of Doctor of Laws from Edinburgh in 1892. He was author of *Sanitary Primer for Indian Schools*, 1879; *Cholera, What Can the State do to Prevent it?*, 1884 and a *Sanitary Primer for Burmese Schools*, 1886.

James Dalton[6]

James Dalton entered the Madras Medical Service as assistant surgeon on 1 July 1791, rising to the rank of surgeon on 14 May 1800 and of superintending surgeon in 1815. He died in England on furlough on 16 September 1823. In addition to his duties at the Native Infirmary, Dalton was involved in the care of the insane in Madras. He was placed in charge of the Madras Lunatic Asylum in 1807 and bought the property from its founder, Valentine Conolly, for 26 000 pagodas.[7] He remained in charge until John Underwood was appointed in December 1814. Dalton rebuilt Conolly's original asylum, situated at Kilpauk, substantially enlarging the accommodation provided. It was known in Madras as 'Dalton's Madhouse'.

Joseph Dougall[8]

Joseph Dougall (1831–79) obtained his Licentiate of the Royal College of Surgeons, Edinburgh, in 1858. He joined the Madras Medical Service as assistant surgeon in 1859, was promoted to surgeon in 1871 and to surgeon-major in 1873. Dougall died at Port Blair despite efforts by the Home Department to find a position for him to continue his experiments with leprosy treatment in the Madras presidency.

William Tilbury Fox[9]

William Tilbury Fox (1836–79) was a brilliant medical student, graduating from University College London as MD in 1857. He became interested in research into skin diseases and published *Skin Diseases of Parasitic Origin* in 1863 and, in 1864, a *Treatise on Skin Diseases*, which developed further in subsequent editions, becoming a standard work in dermatology. In 1864 Fox toured the East with the Earl of Hopetown and on his return wrote pamphlets on the spread of cholera in the East and the dermatology of Egypt. Fox became lecturer on skin diseases at Charing Cross Hospital and physician to the skin department of University College Hospital where he was instrumental in obtaining a well-equipped outpatient department and baths. In addition to clinical and teaching work, Fox was on the *Lancet*'s editorial staff and delivered the 1869–70 Lettsomian Lectures before the Medical Society of London. He re-edited Willan's famous *Atlas of Skin Diseases* in 1875. By the time of his death Fox had become a leading authority in dermatology and had contributed substantially to the development of dermatology as a specialized branch of medicine.

James Lawder[10]

James Lawder (1788–1860) joined the Madras Medical Service as assistant surgeon in 1822 and became a Fellow of the Royal College of Surgeons in 1844, retiring in 1848.

Sir James Ranald Martin[11]

Sir James Ranald Martin (1796–1874) was Company Surgeon in Bengal from 1817 to his retirement in 1840. Returning to London, he served on the Royal Commission of Enquiry into the Sanitary Condition of Large Towns and Populous Districts in England and Wales (1843–45). He was a member of other boards and commissions, including the Royal Commission into the Sanitary State of the Army in India which reported in 1863, and participated in the establishment of the military hospital at Netley. Martin's status as one of Britain's leading authorities on tropical disease and medicine was assured by his revision of James Johnson's work, *The Influence of Tropical Climates on European Constitutions*. Martin was knighted in 1860.

Gavin Milroy[12]

Gavin Milroy (1805–86) studied medicine at the University of Edinburgh and was a founder member of the Hunterian Society of Edinburgh. Milroy gained his Licentiate of the Royal College of Surgeons, Edinburgh in 1824 and worked as a medical officer in the Government Package Service to the West Indies and the Mediterranean. Returning to Britain he was co-editor of the *Medico-Chirurgical Review*. He became an acknowledged authority on epidemiology and was employed on several official commissions and committees. From 1849 to 1850 and 1853 to 1855 he was a superintending medical inspector of the General Board of Health, and in 1852 was sent by the Colonial Office to Jamaica to investigate and report on a cholera epidemic. From 1855 to 1856 he served on the Sanitary Commission in the Crimean War. Milroy was honorary secretary of a committee appointed by the Social Science Association in 1858 to enquire into and report on quarantine. Milroy was a member of the Royal College of Physicians' Leprosy Committee and commented on the college's *Report on Leprosy* of 1867. He was secretary of the Epidemiological Society 1862–64 and president 1864–66, and was awarded a civil list pension in 1871.

William Brooke O'Shaughnessy[13]

William Brooke O'Shaughnessy (1808–89) gained his MD at Edinburgh in 1829 and entered the Bengal Medical Service as assistant surgeon in 1833. He became a Member of the Royal Society in 1843, was promoted to surgeon in 1848 and to surgeon-major in 1859, retiring in 1861. O'Shaughnessy was knighted in 1856. In addition to his medical service, he was director general of Telegraphs from 1852–61. He changed his name to O'Shaughnessy-Brooke in 1861. He was author of a *Manual of Chemistry*, 1837 and published three works in 1841: *Report on Poisoning, Lectures on Galvanic Electricity* and *Bengal Dispensatory*. In 1844 he published in Calcutta *The Bengal Pharmacopœia and General Conspectus of Medical Plants Arranged According to the Natural and Therapeutic Systems*, followed in 1853 by the *Electric Telegraph Manual* and many official reports on the introduction and construction of electric telegraphs in India.

H.H. Risley[14]

H.H. Risley, ICS, was census commissioner for India from October 1899 to September 1902 when W.S. Meyer temporarily succeeded him. E.A. Gait became census commissioner on the 23 January 1903.

William Roxburgh[15]

William Roxburgh (1751–1815) was appointed assistant surgeon at Madras on 28 May 1776 and attained the rank of surgeon four years later. His early years of service were spent at Samulcotta in the Northern Circars where he was superintendent of the pepper plantations. He succeeded Patrick Russell as naturalist at Madras in 1789 and was appointed superintendent of the Calcutta Botanical Gardens in 1793 upon the death of their founder, Colonel Kyd. Roxburgh's principal works were *Plants of the Coast of Coromandel selected from Drawings and Descriptions presented to the . . . East India Company*, London, published in three volumes in 1795, 1802 and 1819; *Hortus Bengalensis; or, a Catalogue of the Plants growing in the Honourable East India Company's Botanic Garden at Calcutta*, Serampore, 1814 and the *Flora Indica*, Serampore, of which an incomplete edition was published posthumously in two volumes in 1820 and 1824 with additions by the botanist, Nathanial Wallich, superintendent of the Calcutta Botanical Gardens from 1816 until 1846. The *Flora Indica* was published in full without Wallich's additions in 1832. The 1832 edition was reprinted in Calcutta in 1874.

Gopal Chunder Roy[16]

Gopal Chunder Roy (1844–87) was educated at Calcutta and Glasgow Universities, gaining his Licentiate in Medicine and Surgery from Calcutta in 1865. He was a lecturer in Nagpur Medical School from 1865 to 1867. He became both Member and Fellow of the Royal College of Surgeons in 1870 and gained his MD from Glasgow in 1871. He entered the Bombay Medical Service as assistant surgeon in 1872, was promoted to surgeon in the following year and to surgeon-major in 1884. Roy was a Christian convert and believed in the superiority of British medicine over traditional Indian medicine.

Rudolf Virchow[17]

Rudolf Virchow (1821–1902) eminent pathologist, sanitarian and medical reformer, became a teacher at the University of Berlin in 1847 and in the same year founded, with Benno Reinhardt, the famous *Archiv für Pathologische Anatomie und Physiologie und für klin ische Medizin* (*Archives for Pathological Anatomy and Physiology, and for Clinical Medicine*), known as Virchow's *Archives*. After Reinhardt's death, Virchow became sole editor. Virchow's book *Cellular Pathology: as based upon Physiological and Pathological Histology*, first published in 1858, was one of the foundation stones of modern medicine. By the 1870s, Virchow's contributions to the conceptual structures of medicine had made him world famous. In 1893 he was invited to give the Croonian lectures in London. In 1897, Virchow presided over the Berlin Conference of Leprologists.

Edward John Waring[18]

Edward John Waring (1819–91) served in Jamaica in the Colonial Medical Service from 1841 to 1842. He became a Member of the Royal College of Surgeons in 1842 and joined the Madras Medical Service as assistant surgeon in 1849, being promoted to surgeon in 1864. Waring became a Fellow of the Royal College of Surgeons in 1864 and retired from the Madras Service in the following year. He gained his MD in 1865 and became a Member of the Royal College of Physicians, London, in 1866 and a Fellow of the College in 1871. He was made Companion of the Order of the Indian Empire in 1881 and was awarded the Burma medal in 1852. Waring was author of *Statistics and Pathology of Abscess of the Liver*, 1854; *Manual of Practical Therapeutics*, 1854, which ran to a fourth edition in 1856; *Notes on some of the Diseases of India*, 1857; *Bazaar Medicines of India*, 1860, the sixth edition published in 1901; *The Tropical Resident at Home*, 1866; *Cottage Hospitals*, 1867; *Pharmacopœia of India*, 1868; *Bibliotheca Therapeutics*, two volumes, 1878–79 and two non-medical texts, *The Hospital Prayerbook*, 1888 and *Precious Jewels for Daily Use*.

Notes

Abbreviations used

Adv Gen	Advocate General
ARSCM	*Annual Report of the Sanitary Commissioner for Madras*
Asst	Assistant
Asyl	Asylum
Bull Hist Med	*Bulletin of the History of Medicine*
CC	Chief Commissioner
Cmmttee, Mon Ch & Nat Inf	Committee of the Monegar Choultry and Native Infirmary
Coll	Collector
CSG	Chief Secretary to Government
CUL	Cambridge University Library
DIG	Deputy Inspector General
Dist Mag	District Magistrate
FINS	Friend in Need Society
G in C	Governor in Council
GG	Governor General
GG in C	Governor General in Council
GoI	Government of India
GS	Government Secretary
H	Home
Hist Recds of Aust Sci	*Historical Records of Australian Science*
ICR	*Indian Census Report*
IMD	Indian Medical Department
IMG	*Indian Medical Gazette*
Ind Ann Med Sci	*Indian Annals of Medical Science, or Half-Yearly Journal of Practical Medicine and Surgery*
Inf	Infirmary
Int J Lep	*International Journal of Leprosy*
JLIC	*Journal of the Leprosy Investigation Committee*
J Trop Med	The *Journal of Tropical Medicine*
L & M	Local and Municipal
M	Municipal
Mad J Med Sci	*Madras Journal of Medical Science*
Mad Med J	*Madras Medical Journal*
Mad P	Madras Public
Mad Qtrly J Med Sci	*Madras Quarterly Journal of Medical Science*
Mad Qtrly Med J	*Madras Quarterly Medical Journal*
MARCD	*Annual Report of the Civil Dispensaries of the Madras Presidency*

MARCHD	*Annual Reports on the Civil Hospitals and Dispensaries of the Madras Presidency*
MAS	*Modern Asian Studies*
MCR	*Madras Census Report*
Med Bd Repts	*Medical Reports Selected by the Medical Board compiled from the Records of their Office, Madras*
MO	Medical Officer
MTamL	*Madras Tamil Lexicon*
MTL	The Mission to Lepers in India and the East
NAI	National Archives of India
Nat Asyl	Native Asylum
Nat Inf	Nat Infirmary
Nat Lep Fund	National Leprosy Fund
NNR (Mad)	*Native Newspaper Reports (Madras)*
OIOC	Oriental and India Office Collections
OP	Official Publications
P	Public
PH	Public Health
RCP	Royal College of Physicians
Res	Resolution
RMTS	*Reports on Medical Topography and Statistics, compiled from the Records of the Medical Board Office, Madras*
SC	Sanitary Commissioner
Sec	Secretary
Sen	Senior
Skt	Sanskrit
Spdt	Superintendent
SRGI	*Selections from the Records of the Government of India*
SRMG	*Selections from the Records of the Madras Government*
Surg-Gen	Surgeon-General
Surg-Maj	Surgeon-Major
Surg	Surgeon
Tam	Tamil
TLM	The Leprosy Mission Archive
TNA	Tamil Nadu Archives
VM	Visiting Member
WIHM	Wellcome Institute for the History of Medicine

Introduction

1 Michel Foucault, *Madness and Civilization: a History of Insanity in the Age of Reason*, New York, 1988, p. 3.
2 John Iliffe, *The African Poor: a History*, Cambridge, 1987, pp. 214, 219.
3 Megan Vaughan, *Curing Their Ills: Colonial Power and African Illness*, Cambridge, 1991, p. 77.

4 Wellesley C. Bailley, *A Glimpse at the Indian Mission Field and Leper Asylums in 1886–87*, London, 1890, p. 24.
5 Medieval science believed that the four humours or body fluids needed to be in balance for good health. Any imbalance resulted in disease.
6 *MCR*, 1891, p. 166.
7 W.R. Rice, 'Measures suggested in regard to the treatment of lepers', 16 August 1893, in *Papers Relating to the Treatment of Leprosy in India from 1887–1895*, [*SRGI*, Home, No. CCCXXXI], Calcutta, 1896 (Hereafter, *Papers*) p. 301.
8 Radhika Ramasubban, 'Imperial Health in British India, 1857–1900', in Roy MacLeod and Milton Lewis (eds), *Disease, Medicine and Empire: Perspectives on Western Medicine and the Experience of European Expansion*, London, 1988, pp. 38–60.
9 M. Harrison, *Public Health in British India: Anglo-Indian Preventive Medicine, 1859–1914*, Cambridge University Press, Cambridge, 1994, pp. 2–3; David Arnold, 'Crisis and Contradiction in India's Public Health', in Dorothy Porter (ed.), *The History of Public Health and the Modern State*, Amsterdam-Atlanta, 1994, p. 353.
10 Roy MacLeod, 'On Visiting the "Moving Metropolis" Reflections on the Architecture of Imperial Science', *Hist Recds of Aust Sci*, 5, 3 (1982) p. 2.
11 Michael Ignatieff, Review of Michel Perot (ed.), *L'Impossible Prison: Recherches sur le systeme penitentiaire au XIXe siecle*, Paris, 1980, in *Social History*, 7, 2 (1982) pp. 228–9, cited in Anand A. Yang, *Crime and Criminality in British India*, Tucson, 1985, pp. 4–5.

1 Indian and British concepts of leprosy and the leprosy sufferer

1 J. Dalton, Surg, Nat Poor Asyl & Inf, to Sir George Barlow, G in C, 21 July 1811, P, 2 August 1811, p. 4259, TNA.
2 W.J. van-Someren, 'A Brief Historical Sketch of the Madras Leper Hospital with some Notices of Leprosy as it is seen in that Institution', *Mad Qtrly J Med Sci*, III (1861) p. 278; Zacharay Gussow, *Leprosy, Racism, and Public Health*, London, 1989, p. 71; Royal College of Physicians, *Report on Leprosy*, London, 1867 (hereafter, RCP, *Report*) p. vii.
3 Olaf K. Skinsnes, 'Notes from the History of Leprosy', *Int J Lep*, 41, 2 (1973) p. 224, quoted in Gussow, p. 71.
4 Francis Day, 'Elephantiasis Græcorum, or Leprosy', *Mad Qtrly J Med Sci*, I (1860), p. 291.
5 Ibid., pp. 286–300; Shaw's report prepared for the Royal College of Physicians is held in the Tamil Nadu Archive at: J. Shaw, Principal IG, IMD, Madras, to A.J. Arbuthnot, CSG, 16 September 1864, P, 29, 7 October 1864. It is reprinted in the Royal College of Physicians, *Report on Leprosy*, London, 1867, pp. 100–7. Subsequent references to Shaw's report will be to its location in the RCP, *Report on Leprosy*.
6 Ibid., p. 101; Day, pp. 292–4.
7 Ibid., p. 295.

8 Ibid., p. 297.

9 RCP, *Report*, p. 100.

10 Ibid., p. 102.

11 Day, pp. 290–1; Gavin Milroy, *Report on Leprosy and Yaws in the West Indies*, London, 1873, p. 93.

12 Tilbury Fox and T. Farquhar, *On Certain Endemic Skin and Other Diseases of India and Hot Climates Generally*, London, 1876, Appendix 1: 'Abstract of the Original Scheme for Obtaining a Better Knowledge of the Endemic Skin Diseases of India' (hereafter, Fox and Farquar, 'Scheme') p. 24.

13 W. Johnston, 'Observations on Leprosy and on its Treatment by Means of Vaporized Carbolic Acid in Union with Watery Vapor', *IMG*, 2 November 1874, p. 287.

14 T.R. Lewis and D.D. Cunningham, *Leprosy in India*, Calcutta, 1877, p. 9; *Lancet*, 23 February 1861, p. 192.

15 *Reports Forwarded by Surgeon-General E.G. Balfour on Certain Forms of Skin Diseases Observed in the Madras Presidency*, as suggested in the pamphlet by T. Fox and T. Farquar etc. (Supplementary Report), Madras, 1875 (hereafter, *Reports, Balfour*) p. 16.

16 Gussow, p. 72.

17 Roderick E. McGrew, *Encyclopedia of Medical History*, New York, 1985, p. 163; H.V. Carter, 'Memorandum on the Prevention of Leprosy by Segregation of the Affected', Bombay General, 5 March 1884, CUL, OP, p. 9; Fox and Farquhar, 'Scheme', p. 24.

18 RCP, *Report on Leprosy*, p. 103.

19 *Leprosy in India: Report of the Leprosy Commission in India, 1890–91*, Calcutta, 1892 (hereafter, *Report, Leprosy Commission*) p. 64.

20 Ibid., p. 309; Fox and Farquhar, 'Scheme', pp. 29–30.

21 RCP, *Report*, 103.

22 Ibid, pp. iv–v.

23 Lewis and Cunningham, pp. 12–13.

24 RCP, *Report*, pp. lxiv, lxvi, lxxiv.

25 *MARCD*, 1855, p. 7; RCP, *Report*, p. 103.

26 *ICR*, 1901, p. 146.

27 *MCR*, 1871, p. 198.

28 Lewis and Cunningham, p. 1; N. Gerald Barrier (ed.), *The Census in British India: New Perspectives*, New Delhi, 1981, p. ix.

29 *MCR*, 1901, p. 110.

30 *MCR*, 1891, p. 164; Lewis and Cunningham, p. 1.

31 *MCR*, 1871, p. 24; *Madras Census*, 1881, Vol. iii, Appendices, p. 54; *ICR*, 1901, p. 147.

32 *Memorandum on the Census of British India of 1871–72*, London, 1875, p. 37; *MCR*, 1891, p. 165.

33 *MCR*, 1871, p. 199; 1891, pp. 165–6.

34 *MCR*, 1901, p. 116; *ICR*, 1901, p. 147; M. Christian, 'Epidemiology', in R.J. Thangaraj (ed.), *A Manual of Leprosy*, 6th edn, New Delhi, 1989, p. 12.

35 *MCR*, 1901, p. 110.

36 *ICR*, 1901, p. 147; *MCR*, 1901, pp. 110–6; 1891, p. 167; *ICR*, 1901, pp. 146–7; Christian, 'Epidemiology', pp. 12–15.

37 *MCR*, 1871, p. 24.
38 *Madras Census*, 1881, Vol. III, *Appendices*, p. 53.
39 *MCR*, 1901, pp. 111, 116.
40 *MCR*, 1911, p. 143.
41 *ICR*, 1931, pp. 262–3.
42 *ICR*, 1901, p. 146.
43 Ibid., p. 148; *Report, Leprosy Commission*, p. 118; *MCR*, 1901, p. 116.
44 *Reports, Balfour*, p. 14.
45 *MCR*, 1901, Subs Tables 4–5, p. 121.
46 *ICR*, 1901, p. 146; Presidency General Hospital: Table No. 2, *RMTS, Presidency Division of the Madras Army, including Fort St George and its Dependencies*, Madras, 1842, pp. 37–8 and 31–2; Res, 9 August 1889, Court of Wards, 1856 in G.O.16, Jdl (Ord), 9 January 1890, TNA.
47 Rev J. Spring, Pres, Cmmttee, FINS, to CSG, 20 June 1838, Mad P, 16, Diary to 17 July 1838, OIOC.
48 *Report, Leprosy Commission*, p. 113.
49 G. Gogerly, *The Pioneers: a Narrative of Facts Connected with Early Christian Missions in Bengal*, London, 1871, pp. 75–6, cited in G.A. Oddie, '"Orientalism" and British Protestant Missionary Constructions of India in the Nineteenth Century', *South Asia*, XVII, 2 (1994) p. 37; Oddie, 'Orientalism', p. 33; *MCR*, 1891, p. 167.
50 *ICR*, 1901, p. 116; *Bengal, Bihar and Orissa and Sikhim, CR, 1911*, p. 424; Bailley, p. 25.
51 S.N. Brody, *The Disease of the Soul: Leprosy in Medieval Literature*, Ithaca, 1974, pp. 11–12, 133; 100–6; 52, 180–2; *Lancet*, III (1824) pp. 149–50.
52 H.V. Carter, 'On the Symptoms and Morbid Anatomy of Leprosy; with Remarks', *Transactions of the Medical and Physical Society of Bombay, No. VIII, New Series, 1862* (Bombay 1865), p. 30.
53 Gussow, pp. 20, 112–13; A. Hilson, Dpty Surg-Gen, IG, et al., 3 March 1890, H (P), 403, A, May 1890, NAI.
54 Gussow, p. 20; MTL, *12th Annual Report for the Year, 1886*, London, pp. 5–7, TLM.
55 Bailley, p. 24.
56 Ibid., pp. 5, 42; Zachary Gussow and George S. Tracy, 'Stigma and the Leprosy Phenomenon: the Social History of a Disease in the Nineteenth and Twentieth Centuries', *Bull Hist Med*, XLIV (1970) p. 446.
57 Asst Col Sec to Sec, GoI, 26 June 1891, H (Med), 71–2, A, August 1891, NAI.
58 Jane Buckingham, 'The "Morbid Mark": the Place of the Leprosy Sufferer in Nineteenth Century Hindu Law', *South Asia*, XX, 1 (1997) pp. 59–61, 69–73.
59 M. Rama Jois, *Legal and Constitutional History of India*, Vol. II, Bombay, 1984, pp. 9, 34; William Hay Macnaghten, *Principles and Precedents of Moohummudan Law*, 2nd edn, Madras, 1860, p. 89; Neil B.E. Baillie, *A Digest of Moohummudan Law . . .*, Part II, London, 1869, p. 102.
60 'Twice-born' or *dvija*: a male from one of the three highest Varna categories – Brahmin, Kshatriya or Vaishya – who has undergone initiation (birth) into the study of Vedic literature, symbolized by investiture with the sacred thread.

61 John D. Mayne, *A Treatise on Hindu Law and Usage*, 9th edn, Madras, 1922, pp. 6, 9; A.C. Burnell, *Introduction to the Daya-Vibhāga*, p. x, cited in J.H. Nelson, *A View of the Hindu Law as Administered by the High Court of Judicature at Madras*, Madras, 1877, pp. 136–7; J.D.M. Derrett, Religion, *Law and the State in India*, London, 1968, pp. 229–30.
62 *Papers*, p. 34.
63 *MCR*, 1911, p. 151; *Papers*, p. 34.
64 Ibid., pp. 25, 34.
65 Ibid., p. 23; *MCR*, 1911, pp. 143, 151.
66 *Papers*, pp. 26, 34–5, 40.
67 *MCR*, 1911, pp. 150–1.
68 For example *RMTS, Northern Hyderabad and Nagpore Divs, Tennasserim Provinces and Eastern Settlements*, Madras, 1844, p. 55, 65–6; General Table Exhibiting the Total Number of Admissions and Deaths in the Madras Army, 1829–38, *RMTS, Pres Divn, Madras Army*, 1842.
69 W. Johnston, 'Observations on Leprosy and on its Treatment by Means of Vaporized Carbolic Acid in Union with Watery Vapour', *IMG*, 1 January 1875, p. 12.
70 *Civil Medical Code Madras*, Madras, 1898, VIII: I, Para 511, p. 103; *Civil Medical Code Madras*, Madras, 1929, Vol. I, 4th edn, XIII, para 337, p. 167.
71 M. Hammick, Mag, Chingleput, to CSG, Jdl, 9 July 1890, and G.O., G.O. 1269, Jdl, 30 July 1890; G.O. 2477–8, Jdl, 11 October 1894, TNA.
72 *Papers*, p. 34.
73 Day, p. 288.
74 *Papers*, p. 41.

2   Patient or prisoner? Leprosy sufferers in British institutional case

1 Michel Foucault, *Madness and Civilisation*, New York, 1988, p. 3. The book was first published in French by Librairie Plon in 1961, followed by an English translation by Richard Howard (New York: Random House, 1965). *Discipline and Punish* was first published in French by Editions Gallimard in 1975, followed by an English translation by Alan Sheridan (Peregrine Books, 1977).
2 S.L. Bhatia, *A History of Medicine*, New Delhi, 1977, pp. 174–5.
3 *Civil Medical Code Madras*, 1898, p. 323; Day, p. 287.
4 Mr Dick, Sen Member, Nat Poor Fund Cmmttee, to Mad P, 6 February 1809, Extr Mad P, 14 March 1809; Extr P Letter from Madras, 6 February 1810, Bds Coll, F/4/347 8113, OIOC.
5 P, 18 August 1807, cited in Henry Davidson Love, *Vestiges of Old Madras, 1640–1800*, 4 Vols, Delhi, 1988, Vol. III, pp. 236, 498; Mgt Cmmttee, Madras Inf & Nat Poor Asyl, to Mad P, 28 February 1810, Extr Mad P, 27 March 1810, Bds Coll, F/4/347 8113, OIOC; R. C. Sherwood, Sec, Nat Poor Asyl & Nat Inf, to CSG, 31 October 1809, P, 7 November 1809, p. 6843, TNA.
6 D. Hill, Surg-Gen, to Mgt Cmmttee, Nat Poor Fund & Nat Inf, 14 June 1811, Extr Mad P, 14 June 1811, Bds Coll, F/4/380 9563; Mr Dick, Sen

Member, Nat Poor Fund, to Mad P, 6 February 1809, Extr Mad P, 14 March 1809, Bds Coll, F/4/347 8113, OIOC.

7 Cmmttee, Nat Inf & Mon Ch, to Sir George Barlow, G in C, 20 March 1813; Cmmttee, Nat Inf & Mon Ch, to John Abercromby, G in C, 16 July 1813; D. Hill, Sec, G in C, to Cmmttee, Nat Inf & Mon Ch, 27 July 1813, Mad P, 27 July 1813, pp. 4349, 4355, 4357, OIOC.

8 Cmmttee, Mon Ch & Nat Inf, to Hugh Elliot, Gov in C, 1 August 1815; Mad P, 11 August 1815, pp. 2155–9, OIOC.

9 GS to Cmmttee, Mon Ch & Nat Inf, 11 August 1815, Mad P, 11 August 1815, pp. 2155–59; I. Braddock, L. Ragava Chitty, VMS for November 1838, to CSG, 30 November 1838, Mad P, 11, 24 December 1838, OIOC.

10 J. Lawder, Surg, Nat Inf, to VM, Mon Ch, 21 February 1839, Mad P, 44, 19 March 1839, OIOC.

11 Rev. J. Spring, Pres, Cmmttee of FINS, to CSG, 20 June 1838, Mad P, 16, Diary to 17 July 1838, OIOC; Res, 17 June 1840, Mad P, 2, Diary to 23 June 1840; Cmmttee, Mon Ch & Nat Inf, to John Lord Elphinstone, G in C, 30 January 1841, Mad P, 18, Diary to 16 February 1841, OIOC.

12 Sec, Med Bd, to CSG, 2 August 1838, Mad P, 37, 14 August 1838; J. Lawder, Surg, Nat Inf, to Spdg Surg, 25 May 1840, Mad P, 1, Diary to 23 June 1840; J. Lawder, Surg, Nat Inf, to Sec, Mon Ch, 9 July 1839, Mad P, 1, Diary to August 1839, OIOC.

13 J. Lawder, Surg, Nat Inf, to Spdg Surg, Pres, 25 May 1840, Mad P, 1, Diary to 23 June 1840 OIOC.

14 Ibid., and Res, 17 June 1840, Mad P, 2, Diary to 23 June 1840, OIOC; *Standing Orders of the Government Leper Hospital, Madras, 1895*, Madras, 1896, p. 10.

15 RCP, *Report*, pp. lxix–lxx, p. 100; Asst Col Sec, to Sec, GoI, 26 June 1891, H (Med), 71–2, B, August 1891, NAI.

16 RCP, *Report*, p. 227.

17 Arnold, *Colonizing the Body: State Medicine and Epidemic Disease in Nineteenth-Century India*, Berkeley, 1993, p. 248; *Fort St George Gazette*, Friday Evening, 1 April 1842, pp. 251–2.

18 John Shortt, 'Account of the Rise and Progress of the Civil Dispensary, Chingleput', *Ind Ann Med Sci*, July, VI (1859) pp. 521–2; J. Shortt, Zillah Surg, Chingleput, to J. Mayer, Dpty IG Hospitals, Madras, 19 March 1864, P, 53, 13 April 1861, TNA; *MARCHD, 1878*, pp. ix, 4.

19 'Statement showing the Number and Cost of Maintenance of Lepers in Poor Houses and Langarkhanas in Tanjore, Madura and Tinnevelley Districts', 21 September 1898, P, 1276, 7 October 1898; Surg-Maj Gen, C. Sibthorpe, Surg-Gen, to CSG, 18 June 1898, P, 1276, 17 October 1898, TNA.

20 'Statement showing the Diseases Prevalent among the Inmates of the Monegar Choultry and Idiot Asylum during the year 1845', 31 December 1845, Mad P, 20, 24 March 1846, OIOC; Surg-Gen, to CSG, 29 September 1891, Jdl, 2270, 4 November 1891, TNA.

21 Robert Clerk, GS, to Cmmttee, Mon Ch & Nat Inf, 2 April 1840, Mad P, 24, Diary to 7 April 1840, OIOC; Cmmttee, Mon Ch & Nat Inf, to Lord Elphinstone, G in C, 23 January 1839, Mad P, 37, 19 February 1839, OIOC.

204 *Notes*

22 *RMTS, Pres Div Madras Army*, p. 67; G.O.1245, Jdl, 19 July 1889, TNA.
23 J. Dalton, Surg, to Spdt Nat Asyl, 16 Aug 1813; 'Return of Lepers in the Native Infirmary, 1812–13', 26 August 1813, Mad P, 3 September 1813, pp. 5001, 5003–5, OIOC; *MCR*, 1911, p. 150.
24 *Papers*, p. 32.
25 P. Thomson, Dist Mag, Tanjore, to Mr J.F. Price, CSG, 3 June 1891 and G.O. in G.O. 1393, Jdl (Ord), 13 July 1891, TNA.
26 E.G. Balfour, Surg-Gen, IMD, to Under-Sec G, 10 July 1874, P, 56, 24 August 1874, TNA; Van-Someren, p. 278.
27 Erving Goffman, *Asylums: Essays on the Social Situation of Mental Patients and Other Inmates*, Peregrine, 1987, p. 11.
28 Love, Vol. III, p. 499; Cmmttee, Mon Ch & Nat Inf, to Hugh Elliot, G in C, 24 April 1816, Extr Mad P, 17 May 1816, Bds Coll F/4/526 12575, OIOC.
29 J. Braddock, Tres & Vice Sec, Mon Ch, to CSG, 31 July 1839; J. Lawder, Surg, Nat Inf, to Sec, Mon Ch, 9 July 1839, Mad P, 1, Diary to August 1839, OIOC.
30 *RMTS, Pres Div, Madras Army*, p. 66; H. King (ed.), The *Madras Manual of Hygiene*, Madras, 1875, p. 214.
31 Michel Foucault, *Discipline and Punish: the Birth of the Prison*, Peregrine, 1979, pp. 200–1.
32 Cmmttee, Mon Ch & Nat Inf, to Hugh Elliot, G in C, 24 April 1816, Extr Mad P, 17 May 1816, Bds Coll, F/4/526 12575, OIOC.
33 Cmmttee, Mon Ch & Nat Inf, to Hugh Elliot, G in C, n.d., Mad P, 20 January 1817, p. 187, OIOC.
34 Waltraud Ernst, 'The Establishment of "Native Lunatic Asylums" in Early Nineteenth Century British India', in G. Jan Meulenbeld and Dominik Wujastyk (eds), *Studies on Indian Medical History*, Groningen, 1987, p. 174.
35 *Standing Orders, Gvt Lep Hosp* p. 18; *Standing Orders of the Government Ophthalmic Hospital, Madras, 1895*, Madras 1895, p. 30.
36 Love, Vol. III, p. 499.
37 Rev. J. Spring, Pres, Cmmttee, FINS, to CSG, 20 June 1838, Mad P, 16, Diary to 17 July 1838; J. Lawder, Surg, Nat Inf, to Spdg Surg, 25 May 1840, Mad P, 1, Diary to 23 June 1840, OIOC.
38 *Ind Med Rec*, 1 August 1890, p. 154.
39 Bailley, p. 37; Day, p. 286; *Standing Orders, Gvt Lep Hosp*, p. 4.
40 Van-Someren, p. 278; *MARCD, 1852*, p. 8; Day, p. 288.
41 *Marcál*: a grain measure in use in the Madras presidency which varied in different localities, though the most usual quantity was 12 seers of grain. From 1846 the standard was fixed at 800 cubic inches. Yule, Henry and A.C. Burnell, Hobson-Jobson, *A Glossary of Colloquial Anglo-Indian Words and Phrases*, Delhi, 1968. p. 567; 1 seer=approx. 1 kilogram or 1 litre; 1 vis= approx. 1.5 kilograms. Ivor Lewis, *Sahibs, Nabobs and Boxwallahs: a Dictionary of the Words of Anglo-India*, Bombay, 1991, pp. 212, 245.
42 Geo Pearse, Sec Med Bd, to CSG, 9 July 1840, Mad P, 35, Diary to 28 August 1840; Res, 27 August 1840, Mad P, 36, Diary to 28 August 1840; 'Expenses', Cmmttee, Mon Ch & Nat Inf, to Barlow, G in C, 20 March

1813, Mad P, 27 July 1813, pp. 4351–3; 'Expenses', Cmmttee, Mon Ch, to Abercromby, G in C, 20 December 1813, Mad P, 4 January 1813, pp. 83–5, OIOC.

43　'Report on Jail Administration in Madras', 20 May 1891, G.O. 2286, Mad Jdl, 6 November 1891, OIOC.

44　*Standing Orders, Gvt Lep Hosp*, p. 11; A.P. Howell, Sec, GoI, H (Med), to CSG, 29 January 1877, P, 74, 15 February 1877; A.P. Howell, Sec, GoI, H (Med), to CSG, 7 May 1877, P, 110, 31 May 1877; CSG, to Sec, GoI, 15 February 1877, P, 75, 15 February 1877, TNA.

45　*Papers*, p. 32; *Standing Orders, Gvt Lep Hosp*, pp. 11–12, 18.

46　Ignatieff, cited in Yang, p. 5.

47　IG Hospitals, IMD, to CSG, 21 March 1873, P, 115, 28 March 1873 TNA.

48　Surg W. Macrae, Surg 1st Dist, to W.H.S. Burn, Dpty Surg-Gen, IMD, Presidency and Nthn Dist, 19 December 1874, P, 36, 13 January 1875, TNA.

49　David Arnold, 'European Orphans and Vagrants in the Nineteenth Century', *Journal of Imperial and Commonwealth History*, January, VII, 2 (1979) p. 82; 'Report on the Police of Madras', 24 February 1817, Mad P, 24 March 1817, p. 1089, OIOC; Van Someren, p. 293; Surg-Maj van-Someren, 1st Dist, to W.H.S. Burn, Dpty Surg-Gen, IMD, Presidency and Nthn Dist, 5 September 1873, P, 49, 17 September 1873, TNA.

50　*Report on the Treatment of Leprosy with Gurjon Oil and Other Remedies in the Hospitals of the Madras Presidency*, [SRMG, No. LII], Madras, 1876, (hereafter, *Report, Gurjon Oil*) p. 60.

51　Van-Someren, p. 274.

52　'Report on the Police of Madras', 24 February 1817, Mad P, 24 March 1817, p. 1089, OIOC.

53　Cmmttee, Mon Ch & Nat Inf, to Hugh Elliot, G in C, n.d., Mad P, 20 January 1817, pp. 186–92, OIOC.

54　'Report on the Police of Madras', 24 February 1817, Mad P, 24 March 1817, p. 1089, OIOC.

55　Cmmttee, Mon Ch & Nat Inf, to Hugh Elliot, G in C, n.d., Mad P, 20 January 1817, p. 191, OIOC.

56　Van-Someren, p. 277.

57　Cmttee, Mon Ch, to Munro, G in C, 22 December 1823, Mad P, 12, 30 December 1823, OIOC.

58　W. Taylor, Spdt, Mon Ch, to VMs, Mon Ch, 18 January 1840, Mad P, 22, Diary to 7 April 1840, OIOC.

59　Major H. Moberly, Sec Milt Bd, to GS, 23 September 1839, Mad P, 20, 29 October 1839, OIOC.

60　Res, 1 October 1839; George Pearse, Sec Med Bd, Med Bd Report, 24 October 1839, Mad P, 20, 29 October 1839; Res, 29 October 1839, Mad P, 21, 29 October 1839, OIOC.

61　Res, 17 June 1840, Mad P, 2, Diary to 23 June 1840, OIOC.

62　J. Lawder, Surg, Leper Hosp, to Spdg Surg, Pres, 3 December 1840, Mad P, 37, 22 December 1840, OIOC.

63　Geo Pearse, Sec, Med Bd, to CSG, 17 December 1840, Mad P, 37, 22 December 1840, OIOC.

64　Van-Someren, p. 275.

65   Geo Pearse, Sec, Med Bd, to CSG, 17 December 1840, Mad P, 37, 22 December 1840, OIOC.
66   Res, 22 December 1840, Mad P, 38, 22 December 1840, OIOC.
67   RCP, *Report*, p. vi; Leonard Rogers and Ernest Muir, *Leprosy*, 2nd edn, London, 1940, p. 54.
68   Van-Someren, p. 275.

3   Colonial medicine in the indigenous context

1   Mel Gorman, 'Introduction of Western Science into Colonial India: Role of the Calcutta Medical College', *Progs Am Phil Soc*, 132, 3 (1988) p. 279, note 10.
2   *Tamil Lexicon*, University of Madras, 1982, Vol. I, (hereafter, *MTamL*) p. 198; Kamil V. Zvelebil, *The Smile of Murugan on Tamil Literature of Southern India*, Leiden, 1973, pp. 225, 220, 228; Kamil V. Zvelebil, *The Poets of the Powers*, London, 1973, pp. 132–3.
3   Kamil V. Zvelebil, *Companion Studies to the History of Tamil Literature*, Leiden, 1992, pp. 234–49, notes 7–8, pp. 237–8; R. Manickavasagam, 'Contribution of Agathiyar to Siddha System of Medicine', in S.V. Subramanian and V.R. Madhavan (eds), *Heritage of the Tamils: Siddha Medicine*, Madras, 1983, p. 592.
4   B.V. Subbarayappa, 'Chemical Practices and Alchemy', in D.M. Bose, S.N. Sen and B.V. Subbarayappa (eds), *A Concise History of Science in India*, New Delhi, 1971, p. 335; Kamil V. Zvelebil, *Tamil Literature*, Wiesbaden, 1974, p. 55; Zvelebil, *The Smile of Murugan*, pp. 220, 228; V. Narayanaswami, 'Ayurveda and Siddha Systems of Medicine – a Comparative Study', in Subramanian and Madhavan, p. 569; Manickavasagam, p. 588; Zvelebil, *Companion Studies*, p. 238.
5   Whitelaw Ainslie, *Materia Indica*, 2 Vols, London, 1826, (Vol. I, Delhi, 1986 Vol. II, Delhi, 1984), p. xxx.
6   Narayanaswami, p. 571. Francis Zimmermann, *The Jungle and the Aroma of Meats: an Ecological Theme in Hindu Medicine*, London, 1987, *passim*; Ainslie, Vol. II, pp. xiii–xiv.
7   Julius Jolly, *Indian Medicine*, 2nd edn, New Delhi, 1977 p. 51; C.D. Maclean (ed.), *Manual of the Administration of the Madras Presidency*, 3 Vols, 1885–93 (New Delhi 1987–90), Vol. III *Glossary of the Madras Presidency*, Madras 1893 (New Delhi, 1990), p. 955; *RMTS, Sthn Div, Madras Army*, Madras, 1843, p. 146; Ainslie, Vol. II, pp. xxx–xxxi; E. Valentine Daniel, 'The Pulse as an Icon in Siddha Medicine', in E. Valentine Daniel and Judy F. Pugh (eds), *South Asian Systems of Healing, Contributions to Asian Studies*, Vol. XVIII, p. 115. Thanks to Dr Geoffrey Samuel for this reference; Narayanaswami, pp. 574–5.
8   Ibid., pp. 569–70; Zvelebil, *The Smile of Murugan*, p. 223; Subbarayappa, pp. 335–8; D. Andiappa Pillai, 'Kalluppu in Muppu', in Subramanian and Madhavan (eds), p. 147.
9   K.R. Krishnan, 'Siddha Medicine during the Period of the Marattais' [sic] in Subramanian and Madhavan (eds), pp. 55–6; Maclean, *Glossary*, p. 955; Ainslie, Vol. II, pp. xiii–iv.

10 Poonam Bala, *Imperialism and Medicine in Bengal: a Socio-Historical Perspective*, New Delhi, 1991, p. 146; *The Ordinances of Manu*, trans A.C. Burnell, Edward W. Hopkins (ed.), London, 1884 (2nd edn, New Delhi, 1971), x. 8, 47; Jolly, p. 26.

11 Maclean, *Glossary*, p. 955; Jolly, p. 22; C.K. Sampath, 'Evolution and Development of Siddha Medicine', in Subramanian and Madhavan (eds), p. 14; Krishnan, p. 58; *RMTS, Ceded Districts*, Madras, 1844, p. 65; Charles Leslie, 'The Ambiguities of Medical Revivalism in Modern India', in C. Leslie (ed.), *Asian Medical Systems: a Conference Study*, Los Angeles, 1976, pp. 358–9; Ainslie, Vol. II, pp. ix–x.

12 A. Suresh, G. Veluchamy, 'Surgery in Siddha System', in Subramanian and Madhavan (eds), pp. 468, 462; *RMTS Sthn Div, Madras Army*, 1843, p. 106; Jolly, pp. 37–41, 55; Manickavasagam, p. 594.

13 Arnold, *Colonizing the Body*, pp. 45, 53; James Moores Ball, *The Body Snatchers*, New York, 1989, p. 122; Ruth Richardson, *Death, Dissection and the Destitute*, Penguin, 1988, pp. 131–44, 160.

14 *RMTS Sthn Div, Madras Army*, 1843, p. 106; Maclean, *Glossary*, p. 955; Ainslie, Vol. II, p. ix; O.P. Jaggi, *Indian System of Medicine*, History of Science and Technology in India, Vol. IV, Delhi, 173, pp. 154–6; 175.

15 Maclean, *Glossary*, p. 955; Harrison, pp. 40–1; Jolly, pp. 22–3.

16 K.N. Kuppuccāmi Mutaliyār, *Citta maruttuvam*, Part 1, Madras, 1954, pp. 529–36; K.C. Uttamarāyan, *Cittar aṟuvai maruttuvam*, Madras, 1968, p. 78; Jolly, p. 117; Yūkimuṉivar, *Vaittiya cintāmaṇi, (Perunūl 800)*, Part I, Madras, 1976, p. 356; *Carapēntira vaitiya muṟaikaḷ*, pp. xxii–xxvii.

17 *Reports, Balfour*, pp. 9–15.

18 *Citta maruttuvam*, pp. 530–36.

19 Akattiyar, *Kaṉmakāṇtam 300*, Madras, 1976, pp. 12, 16; Jolly, pp. 117, 119; *Citta maruttuvam*, p. 529; *Vaittiya cintāmaṇi*, pp. 356–7.

20 Jolly, p. 15; Ainslie, Vol. II, p. x; Narayanaswami, p. 568; Maclean, *Glossary*, p. 955; *Kaṉmakāṇtam*, p. 12.

21 Ainslie, Vol. I, p. x; Vol. II, pp. v, xxxii–xxxv; Both Arnold and Harrison also draw attention to this point. *Colonizing the Body*, p. 45; Harrison, p. 41.

22 *RMTS, Coorg*, Madras, 1844, p. 23.

23 *RMTS, Mysore Div, Madras Army*, Madras, 1844, p. 14.

24 J.R. Martin in *Progs of the Governors of the Native Hospital Relative to the Establishment of the Fever Hospital*, Appendix C, Calcutta, 1838, p. 59, cited in Arnold, 'Medical Priorities and Practice in Nineteenth-Century British India', *South Asia Research*, 5, 2 (1985) p. 176.

25 Arnold, *Colonizing the Body*, p. 47; Crawford, *A History of the Indian Medical Service 1600–1913*, Vol. I, London, 1914, p. 22, cited in David Arnold (ed.), *Imperial Medicine and Indigenous Societies*, Delhi, 1989, p. 11.

26 P.D. Gaitonde, *Portuguese Pioneers in India: Spotlight on Medicine*, Bombay, 1983, p. 109; T.J.S. Patterson, 'The Relationship of Indian and European Practitioners of Medicine from the Sixteenth Century', in Meulenbeld and Wujastyk (eds), pp. 119–23; Arnold (ed.), *Imperial Medicine and Indigenous Societies*, p. 11; Harrison, p. 40.

27 Patterson, pp. 127–8; Bala, pp. 41–4; Harrison, p. 41.

28 Arnold, *Colonizing the Body*, p. 13; John Chandler Hume, 'Colonialism and Sanitary Medicine: the Development of Preventive Health Policy in the Punjab, 1860 to 1900', *MAS*, 20, 4 (1986) p. 703; Bala, p. 47; Patterson, p. 128;.
29 W.R. Cornish, Surg-Gen, Madras, to R. Davidson, CSG, P, 21 August 1880, *MARCHD 1879*, pp. 8–9.
30 *Civil Medical Code, Madras*, 1898, p. 116.
31 Arnold, *Colonizing the Body*, pp. 45, 47; *Progs of the Committee on Indian Drugs, Med Bd Repts*, 1855, pp. 418–9.
32 Ainslie, Vol. I., pp. x–xi.
33 Edward John Waring, *Pharmacopœia of India*, London, 1868, p. vi; *Lancet*, 1 April 1865, p. 353; Arnold, *Colonizing the Body*, p. 47.
34 Ainslie, Vol. I, p. x; Waring, pp. v–x.
35 Arnold, *Colonizing the Body*, pp. 54–6.

4 Leprosy treatment: Indigenous and British approaches

1 *Citta maruttuvam*, Part I, p. 50.
2 J. Dalton, Surg, to Spdt, Nat Asyl (Infirmary), 16 August 1813, Mad P, 3 September 1813, p. 5002, OIOC.
3 Van-Someren, pp. 286, 288–9; *Report, Gurjon Oil*, pp. 60–1; *Citta maruttuvam*, pp. 84, 539–43; *Akattiyar paḷḷu*, stanzas 11–15; Carapēntira vaitiya muṟaikaḷ, pp. 154–8, 163–4, 171.
4 Van-Someren, p. 288.
5 Patterson, p. 128.
6 *Carapēntira vaitiya muṟaikaḷ*, Tanjore, 1965, pp. 5, 171, 154–5; *Akattiyar paḷḷu*, stanzas 11–13.
7 Van-Someren, pp. 286, 288–9.
8 Fátima da Silva Gracias, *Health and Hygiene in Colonial Goa*, 1510–1961, New Delhi, 1994, p. 170.
9 Ainslie, Vol. II, pp. 454–5; Waring, pp. 53–5, 444; J.P. Grant, *Mad Med J*, I (1839) p. 420; Dr D. Dunbar, *Mad Med J*, IV (1842) p. 80, cited in Waring, p. 444.
10 *Kaṉmakāṇtam*, pp. 12–13.
11 *Carapēntira vaitiya muṟaikaḷ*, pp. 165–6, 168–9; Cittar a ṟuvai maruttuvam, p. 84.
12 Arnold, *Colonizing the Body*, pp. 47–8; George Playfair, *The Taleef Shereef, or Indian Materia Medica*, [Calcutta 1833], Dehra Dun, 1984, pp. 170–2; Muhammed Abdulla Sahib, *Ūnāni vaittiya thathu virthi bodhini*, Part II, Madras, 1893, pp. 7, 174.
13 Van-Someren, pp. 285–6.
14 J. Mitchell Bruce, *Materia Medica and Therapeutics*, 3rd edn, London, 1886, p. 77; Waring, pp. 343, 112, 191, 278; F.F. Cartwright, 'Antiseptic Surgery' in F.N.L. Poynter (ed.), *Medicine and Science in the 1860s . . .*, London, 1968, p. 79; J. Forbes Royle, *A Manual of Materia Medica and Therapeutics*, 3rd edn, (revised F.W. Headland), London, 1856, pp. 172, 400–2, 225.

15 GS to Cmttee, Mon Ch & Nat Inf, 27 July 1813, Mad P, 27 July 1813, p. 4357, OIOC; Van-Someren, p. 286; J. Dalton, Surg, to Spdt, Nat Asyl, 16 August 1813, Mad P, 3 September 1813, pp. 5000–1, OIOC.

16 *Lancet*, 9 April 1864, pp. 408–9; 4 August 1866, p. 142; Van-Someren, p. 288.

17 *Lancet*, 25 August 1860, p. 188.

18 Van-Someren, pp. 287–9.

19 *Akattiyar paḷḷu*, st. 11–13.

20 *Cittar aṟuvai maruttuvam*, pp. 84–5. *Akattiyar paḷḷu*, st. 14–15.

21 W. Johnston, 1 January 1875, pp. 9–10; Cecilia E. Mettler, *History of Medicine*, Philadelphia, 1947, p. 651; A. Hunter, 'Report Upon the Hydrocotyle Asiatica', *Med Bd Repts*, Madras, 1855, pp. 356–7.

22 *IMG*, 1 December 1869, p. 272; Waring, pp. 269–70.

23 J.M. Fleming, 'On the Curability of Leprosy', *IMG*, 1 June 1871, pp. 119–20; 1 March 1871, pp. 53–4; 2 January 1871, p. 14; *Lancet*, 3 June 1871, p. 756; 18 March 1871, p. 378.

24 *Report, Gurjon Oil*, pp. 60–1, 82–4, 40.

25 Johnston, 1 January 1875, pp. 11–12.

26 Arnold, *Colonizing the Body*, p. 251; W.R. Cornish, Surg-Gen, Madras, to C.G. Master, CSG, P, 30 June 1881, *MARCHD*, 1880, pp. 8–9.

27 Surg-Maj Edward A. Birch, 'Gleanings from a Mofussil Practice', *IMG*, 1 July 1878, p. 182; March 1883, p. 83.

28 E. Lawrie, 'Nerve Stretching in Anæsthetic Leprosy', *IMG*, 2 September 1878, p. 229, note; 1 October 1878, p. 270.

29 J.R. Wallace, 'Nerve-Stretching in Anæsthetic Leprosy', *IMG*, 1 November 1880, p. 301.

30 Ibid., p. 301; 'Cases reported by Assistant-Surgeon, Mohendra Nath Ohdedar', *IMG*, 1 August 1882, p. 213; G.C. Roy, 'Some Remarks on Leprosy', *IMG*, 1 February 1881, p. 46.

31 W.R. Cornish, Surg-Gen, Madras to C.G. Master, CSG, P, 30 June 1881, MARCHD 1880, p. 8. Arnold makes the same point in *Colonizing the Body*, p. 251.

32 R.H. Bakewell, MO, Health, to Trinidad, 'Rules for the Treatment of Lepers by Dr Beauperthuy's Method', in *Correspondence Relating to the Discovery of an Alleged Cure of Leprosy*, G.B., PP, Cmmns, 1871, c. 362, Vol. LVI (hereafter, *Correspondence*) pp. 41–2.

33 J. Dougall, *Report on the Treatment of Leprosy with Gurjon Oil with Other Papers; Extracted from the 'Indian Medical Gazette'*, Calcutta, 1876, Tracts, 516, pp. 1, 22–3, OIOC; D.J. Crawford, *A History of the Indian Medical Service, 1600–1913*, 2 Vols, London 1914, Vol. II, p. 318.

34 Maclean, *Glossary*, pp. 69, 1024, 336; Waring, p. 32; J.C. Ghose, *A Monograph on Chaulmoogra Oil and its Use in the Treatment of Leprosy as Explained in the Ayurveda*, Madras, 1917, pp. 11–12.

35 Bala, pp. 48–9; Waring, pp. 63–4, 441; Dougall, pp. 22–3.

36 Maclean, *Glossary*, p. 1024; *Report, Gurjon Oil*, p. 23.

37 Dougall, pp. 1–2.

38 Ibid., p. 5.

39 Manu, v. 125–6.

40 Dougall, pp. 15–18.

41 Waring, pp. 26, 440; F.J. Mouat, 'Notes on Native Remedies', *Ind Ann Med Sci*, October 1853 and April 1854, I, 1854, pp. 647–8, 652; Ghose, pp. 4–5, 14.
42 Joseph F. Rock, 'The Chaulmoogra Tree and Some Related Species: a Survey Conducted in Siam, Burma, Assam, and Bengal', US Dept Agriculture, Bull No. 1057, Washington, D.C., Professional Paper, 24 April, 1922, pp. 3–6; John R. Trautman, 'The History of Leprosy', in Robert C. Hastings (ed.), *Leprosy*, 2nd edn, London, 1994, p. 17.
43 Mouat, p. 647; Waring, p. 26; Maclean, *Glossary*, p. 159.
44 Leonard Rogers, 'The Successful Treatment of Leprosy', *the Practitioner*, August, CVII, 2 (1921) pp. 79–80; Maclean, *Glossary*, p. 160; Ghose, pp. 4–7.
45 Maclean, *Glossary*, pp. 160, 474; *ARSCM*, 1875, Appendix 1, 'Leprosy', p. xv.
46 Ainslie, Vol. II, pp. 235–6; Waring, p. 27.
47 E.G. Balfour, Surg-Gen, IMD, to Lieut Col J. Goddard, Dist Engineer, Presidency, 31 July 1874, P, 57, 24 August 1874; E.G. Balfour, Surg-Gen, IMD, to CSG, 19 November 1874, P, 26–28, 8 December 1874, TNA; *ARSCM*, 1875, Appendix I, 'Leprosy', pp. xv–xviii.
48 *Report, Leprosy Commission*, p. 338; Surg-Capt. F.J. Crawford, *Syllabus of the Course of Lectures on Materia Medica and Their Therapeutics in the Madras Medical College*, Madras 1895, pp. 6, 9; *Syllabus*, 1897, p. 9; *Syllabus*, 1898, p. 8; *Civil Medical Code Madras*, 1929, Vol. I, Sn xv, para 537, pp. 231–3.
49 *Report, Leprosy Commission*, pp. 338–40; R. Ganapati, 'Therapy of Leprosy', in Thangaraj (ed.), pp. 144–5; K.K. Gupta, 'An Easy Method of Keeping Hydnocarpus Oil Constantly Hot During Injection', *Leprosy in India*, July, 1934, p. 136.
50 Rogers and Muir, p. 232.
51 P.C. Banerjea to the Editor, *IMG*, 1 June 1877, p. 166.
52 Waring, p. 107.
53 Ghose, pp. 8, 10; Mouat, p. 652; Maclean, *Glossary*, pp. 160, 577.
54 Arnold, *Colonizing the Body*, pp. 59–60.
55 M.N. Pearson, 'The Thin Edge of the Wedge: Medical Relativities as a Paradigm of Early Modern Indian-European Relations', *MAS*, Vol. 29, No. 1, 1995, p. 170; *RMTS, Sthn Div, Madras Army*, 1843, p. 146; Ghose, p. 13.
56 M.C. Koman, to Surg-Gen, Madras, 31 December 1919, Madras, L & M (Pub H), 125, 11 February 1921, OIOC; N. Kandaswamy Pillai, *History of Siddha Medicine*, Madras, 1979, pp. 560–7.
57 David Arnold, 'Smallpox and Colonial Medicine in Nineteenth Century India', in Arnold (ed.), *Imperial Medicine and Indigenous Societies*, p. 53; F.A. Marglin, 'Small Pox in Two Systems of Knowledge', in F.A. and S.A. Marglin (eds), *Dominating Knowledge: Development Culture and Resistance*, Oxford, 1990, pp. 115–6, 118.
58 *MARCHD*, 1880, p. 16.
59 *RMTS, Centre Div, Madras Army*, Madras, 1843, p. 9.
60 Arnold, *Colonizing the Body*, pp. 249–50.
61 Surg, George Pearse, Sec, Med Bd, to I.F. Thomas, CSG, 20 September 1845, Mad P, 7, 7 October 1845, OIOC.
62 *Report, Gurjon Oil*, p. 5; Day, p. 288; Dougall, pp. 18–19.
63 *Report, Gurjon Oil*, pp. 6–7, 67.

64 Ibid., p. 40.

65 Ibid., p. 23; J. Dalton, Surg, to Spdt, Nat Asyl, 16 August 1813, Mad P, 3 September 1813, p. 5001, OIOC.

66 *Report, Gurjon Oil*, pp. 6–7, 25; *RMTS, Coorg*, 1844, pp. 23–4.

67 Surg-Maj W.J. van-Someren, 1st Dist, to DIG, IMD, Presidency, 10 March 1873, H [Med] 74, A, May 1876, NAI; *Report, Gurjon Oil*, p. 62.

68 *MARCD, Official Year, 1876–77*, pp. 71, 104.

69 Rock, pp. 3–5; Robert R. Jacobson, 'Treatment of Leprosy', in Robert C. Hastings (ed.), *Leprosy*, 2nd edn, London, 1994, p. 317; Cyril Davey, *Caring Comes First: the Leprosy Mission Story*, Basingstoke, 1987, p. 60.

70 H. Krishna Sastri, *South Indian Images of Gods and Goddesses* (Madras 1916), New Delhi, 1986, pp. 248, 251, 227; Monier Williams, *Religious Thought and Life in India* (London 1883), New Delhi, 1974, p. 324; Bhagavat Singh Jee, *A Short History of Aryan Medical Science*, London, 1896, p. 33; Henry Whitehead, *The Village Gods of South India* (1921), Madras, 1988, pp. 31–2, 37, 48, 54–56.

71 J. Dalton, Surg, to Spdt, Nat Asyl, 16 August 1813, Mad P, 3 September 1813, p. 5002, OIOC.

72 *Report, Gurjon Oil*, pp. 82–4.

73 Arnold, 'Small Pox and Colonial Medicine', p. 49; Arnold, *Colonizing the Body*, p. 143; Arthur Neve, 'Leprosy and Christianity', *J Trop Med*, 15 May, VIII (1905) pp. 146–7.

74 *Papers*, pp. 16, 87.

75 Ibid., p. 33; Day, p. 286; *MARCHD*, 1879, p. 78; Bailley, p. 37; *Report, Gurjon Oil*, pp. 82–84.

76 Arnold, *Colonizing the Body*, pp. 252–3; Bala, p. 109; *MARCHD*, 1879, pp. 8–9.

5 Leprosy research and the development of colonial medical science

1 Arnold, *Colonizing the Body*, pp. 16–17.

2 George Basalla, 'The Spread of Western Science', *Science*, 156, 1967, pp. 613, 611.

3 Arnold, *Colonizing the Body*, p. 14.

4 *Report on Leprosy and Yaws in the West Indies*, p. 94.

5 D.G. Crawford, *Roll of the Indian Medical Service, 1615–1930*, London, 1930, pp. 181, 338, 468; Dougall, p. 1; Harrison, pp. 19, 21, 27, 15–16, 30–2.

6 Carter, 'Morbid Anatomy', p. 1.

7 Dougall, p. 11; H. D. Cook, Surg, 1st Dist, in charge, Madras Leper Hosp, to Sec to Surg-Gen, Madras, 22 June 1888; G. Bidie, Surg-Gen, to CSG, 16 December 1888, P, (Ord), 137, 11 February 1889, TNA.

8 Maj-Gen D.M. Stewart, CC, Andaman and Nicobar Islands, Spdt Port Blair, to Sec, GoI, 14 December 1874, P, 61, 12 August 1875, TNA; *Correspondence*, p. 40.

9 J. Dalton, Surg, Nat Poor Asyl & Inf, to Sir George Barlow, G in C, 21 July 1811, P, 2 August 1811, p. 4260, TNA.

10 Van-Someren, pp. 285–7; David Hill, Sec, G in C, to Cmttee, Mon Ch & Nat Inf, 27 July 1813, Mad P, 27 July 1813, p. 4357, OIOC.
11 W.F. Bynum, *Science and the Practice of Medicine in the Nineteenth Century*, Cambridge, 1994, p. 109.
12 K. Radhakrishnan 'Siddha Medicine for the Skin Diseases', in Subramanian and Madhavan (eds), p. 413; G. Geetha, 'Kaya Kalpa Mooligai in Siddha Medicine', in ibid., pp. 135, 143–5; MTamL, Vol. II, p. 877.
13 Ainslie, Vol. II, pp. 473–4; Hunter, pp. 356–66.
14 *Med Bd Repts*, p. 375; *MARCHD, 1854*, p. 7; Waring, pp. 107, 448.
15 Hunter, pp. 366, 370.
16 Harrison, pp. 150, 156, 158.
17 Arnold, 'Medical Priorities', p. 171.
18 Ruth G. Hodgkinson, 'Social Medicine and the Growth of Statistical Information', in Poynter (ed.), pp. 183–5.
19 Arnold, 'Medical Priorities', p. 171; Maclean, *Manual*, Vol. I, pp. 512–3, note; Harrison, p. 78.
20 'The Edinburgh Medical Journal on "Sanitary Science" and Medical "Statistics"', *Mad Qtrly J Med Sci*, II (1861) p. 197; *Lancet*, 27 May 1865, pp. 572–3.
21 Arnold, 'Medical Priorities', pp. 168, 171; Maclean, *Manual*, Vol. I, pp. 512–3, note; *ARSCM*, 1875, Appendix 1, 'Leprosy', p. iii.
22 *Lancet*, 25 May 1867, p. 631.
23 Crawford, *Indian Medical Service*, II, pp. 452–57.
24 F.B. Smith, *The People's Health, 1830–1910*, London, 1990, pp. 11, 296, 346; Peterson, p. 38; Harrison, pp. 9–10; Bala, pp. 65–70.
25 *Mad Qtrly Med J*, I (1839) pp. iii–iv, viii.
26 J.D.M. Derrett, 'J.H. Nelson: a Forgotten Administrator-Historian of India', in C.H. Philips (ed.), *Historians of India, Pakistan and Ceylon*, London, 1961, p. 356; Harrison, p. 13; 'Prospectus', *Mad J Med Sci*, I (1852) p. 1.
27 *Mad Qtrly Med J*, I (1839) pp. ix–xii; *Mad Qtrly J Med Sci*, I (1860) pp. ii–vi.
28 Harrison, p. 25.
29 *Mad Qtrly J Med Sci*, V (1862) pp. 187–9.
30 Carter, 'Morbid Anatomy', p. 1; H.V. Carter to Dr H. Pitman, Sec, RCP, London, 15 January 1863, RCP, *Report on Leprosy*, p. 225.
31 *Brit and For Med-Chir Rev*, January, V, IX, (1850) pp. 171–82; cited in Carter, 'Morbid Anatomy', p. 79.
32 Carter, 'Morbid Anatomy', pp. 1–2; Day, p. 288; Arnold, *Colonizing the Body*, p. 108.
33 RCP, *Report on Leprosy*, p. 106; 'A memorandum of points to be studied chiefly with the help of the microscope, in post-mortem examinations of leprosy', *IMG*, 2 November 1868, pp. 251–2.
34 Carter, 'Morbid Anatomy', pp. 23, 80.
35 'The History of Leprosy', *Mad Qtrly J Med Sci*, II (1861) pp. 192–5; Mettler, p. 680.
36 *Lancet*, 28 July 1860, p. 90; 4 August 1860, p. 120; 'The History of Leprosy', pp. 192–3.
37 RCP, *Report*, pp. iii.
38 Minutes, RCP Meeting, 14 June 1862, RCP, *Report*, pp. iv, lxxvi.
39 Ramasubban, 'Imperial Health in British India', p. 43; Hodgkinson, p. 187.

40 'The Edinburgh Medical Journal on Sanitary Science and Medical "Statistics"', *Mad Qtrly J Med Sci*, II (1861) pp. 197–9.
41 Henry A. Pitman, Registrar, RCP, to F. Rogers, Under-Sec State, Colonies, 9 August 1862, P, 61, 17 January 1863, TNA; Circular to Governor of Colonies, 28 August 1862, Old Sanitary Home Correspondence, L/E/Z/90/2, OIOC. I am indebted to Philippa Levine for this reference.
42 *Lancet*, 12 January 1867, p. 63.
43 *Lancet*, 25 May 1867, p. 631; RCP, *Report on Leprosy*, p. xv.
44 A.M. Cooke, *A History of the Royal College of Physicians of London*, Oxford, 1972, Vol. III, p. 826; RCP, *Report on Leprosy*, p. vii.
45 *Lancet*, 5 January 1867, p. 17; 8 October 1864, p. 420; 4 July 1863, p. 25.
46 Arnold, *Colonizing the Body*, pp. 25–6; Harrison, p. 101; Radhika Ramasubban, *Public Health and Medical Research in India: Their Origins Under the Impact of British Policy*, SAREC, Stockholm, 1982, pp. 11–12.
47 Ramasubban, 'Imperial Health in British India', pp. 78–82, 101; Donald A. MacKenzie, *Statistics in Britain, 1865–1930: the Social Construction of Scientific Knowledge*, Edinburgh, 1981, pp. 7–8 and *passim*.
48 Harrison, p. 101, 229, 82; Arnold, 'Smallpox and Colonial Medicine', p. 52.
49 RCP, *Report*, pp. lxxii–lxxiii, 3.
50 MacLeod, 'On Visiting the "Moving Metropolis"', p. 2.
51 RCP, *Report*, p. lxxiv.
52 Gussow, pp. 74, 82, 239, note 64.
53 Day, p. 295.
54 *Lancet*, 29 December 1866, pp. 732–3.
55 Peterson, pp. 38, 6–9, 12.
56 *Lancet*, 12 January 1867, p. 63; 23 February 1867, p. 247.
57 Fox and Farquhar, *Endemic Skin and Other Diseases*, p. vi.
58 Fox and Farquhar, 'Scheme' pp. 1–2, 24.
59 Fox and Farquhar, *Endemic Skin and Other Diseases*, pp. v–vi.
60 *Reports, Balfour*, pp. 1, 8–9.
61 Ibid., pp. 12–13.
62 H.H. Wilson, 'Kushta, or Leprosy: as Known to the Hindus', *Trans Med & Phl Soc Calc*, 1, 3 May 1823, pp. 2–3.

6 The politics of leprosy control

1 *Correspondence*, pp. 1–2.
2 Ibid., p. 9.
3 Cooke, p. 828.
4 *Correspondence*, pp. 9–10.
5 Ibid., p. 39.
6 *Lancet*, 5 June 1875, p. 812; *Correspondence*, pp. 16–19.
7 Ibid., pp. 42–3.
8 *Reports, Balfour*, p. 64.
9 *Lancet*, 1 July 1871, p. 36; *Correspondence*, pp. 44, 43; *Lancet*, 29 April 1871, pp. 586–7.
10 *Correspondence*, pp. 41–2.
11 Ibid., p. 43.

12 *Lancet*, 6 September 1873, p. 340; Smith, pp. 296, 298, 366–71.
13 E.G. Balfour, IG Hospitals, IMD, to W. Huddleston, CSG, Madras, 12 June 1873, H (Med), 74, A, May 1876, NAI.
14 GoI to Sec State, India, 18 May 1876, H (Med), 89, A, May 1876, NAI.
15 Dougall, pp. 34, 46–9.
16 E.G. Balfour, Surg-Gen, IMD, to Lieut Col J. Goddard, Dist Eng, Pres, 31 July 1874, P, 57, 24 August 1874; E.G. Balfour, Surg-Gen, IMD, to CSG, (P), 19 November 1874, P, 26, 8 December 1874, TNA.
17 J.M. Cuningham, SC, GoI, note, 5 January 1875; E.C. Bayley, note, 18 January 1875; A.P. Howell, Sec, GoI, to Surg-Gen, IMD, 5 March 1875; GoI to Sec State, India, 5 March 1875, H (Med), 10–14, A, March 1875, NAI.
18 The debate is collected in Dougall, pp. 1–38, 44–50.
19 Ibid., pp. 38–43.
20 *Lancet*, 13 June 1874, pp. 846–7; 27 June 1874, p. 917; 19 December 1874, pp. 887–8.
21 Dougall, pp. 51–73.
22 'Observations'; Surg-Maj Dougall to CC Andaman and Nicobars, 12 December 1876, P, 18, 3 March 1877; G.O.19, P, 13 March 1877, TNA.
23 J.M. Cuningham, 2 June 1879, note, H, Rev and Agricl (Med), 51–64, A, July 1879, p. 3, NAI.
24 C.B., 5 June 1879, note, H, Rev and Agricl (Med), 51–64, A, July 1879, NAI; John C. Lucas, 29 September 1882, *IMG*, 2 January 1882, pp. 23–4.
25 Johnston, 1 January 1875, pp. 9–10.
26 Review of William Johnston, *Observations on Leprosy, IMG*, 1 September 1879, p. 268; E. Balfour, Surg-Gen, IMD, to CSG, P, 5 January 1874, P, 85–6, 24 January 1874, TNA; *Report, Gurjon Oil*, p. 82.
27 Ibid., pp. 60–1.
28 E.G. Balfour, Surg-Gen, IMD, to Lieut Col J. Goddard, Dist Eng, Pres, 31 July 1874, P, 57, 24 August 1874; E.G. Balfour, Surg-Gen, IMD, to CSG, 19 November 1874, P, 26–8, 8 December 1874, TNA; *ARSCM*, 1875, Appendix 1, 'Leprosy', pp. xv–xvii.
29 Rogers and Muir, p. 103; Gussow, pp. 69, 72–3, 77.
30 H.V. Carter, 'Further Observation on the Prevention of Leprosy by Segregation of the affected', Bombay General, 29 June 1887, p. 2. CUL, OP.
31 Gussow, pp. 78–80.
32 RCP, *Report*, p. 225; H.V. Carter to Ed, 23 January 1863, *Lancet*, 28 February 1863, p. 24.
33 H.V. Carter, to Under-Sec State, India, 23 April, 1874; George Burrows, Pres, RCP, to H.V. Carter, 22 April 1874, Encl Stats & Comm, Despatch to India, 21, 6 August 1874, V/6/302, OIOC.
34 'Remarks of the Army Sanitary Commission on a Report on Leprosy and Leper Asylums in Norway, with reference to India, by Dr Vandyke Carter', 30 June 1874 in Ibid.
35 J.M. Cuningham, SC, GoI, 5 January 1875, note, H (Med), 10–14, A, March 1875; Res, H (Med), 12, A, March 1875, NAI.
36 Harrison, pp. 104, 21–22.
37 *Lancet*, 11 December 1869, p. 813; Harrison, p. 79; *Lancet*, 6 November 1869, p. 652.
38 E.C. Bayley, Sec, GoI, 6 January 1876, note, H (Med), 57–60, A, February 1876, NAI.

39 J.M. Cuningham, SC, GoI, 5 April 1875, note, H (Med), 18–22, A, April 1875, NAI.

40 J.M. Cuningham, SC, GoI, 18 January 1876, note, H (Med), 57–60, A, February 1876, NAI.

41 Mettler, p. 638; 'The Memorial of H.V. Carter, to Kimberley, Sec State, India, 18 August, 1884', p. 2, Carter Papers, *Western Manuscripts*, 5809–5826, 2/1, WIHM; Fox and Farquhar, *Endemic Skin and Other Diseases*, pp. 278–85.

42 Arnold, *Colonizing the Body*, p. 192; Hume, pp. 709–17; *Lancet*, 25 September 1875, pp. 458–9.

43 Harrison, p. 109; *IMG*, 1 May 1875, pp. 129–32; GoI to Sec State, India, 14 July 1879, H, Rev & Agricl (Med), 64, A, July 1879, NAI.

44 Carter, 'Memorial', p. 2.

45 *Papers*, pp. 11, 18.

46 Ibid., pp. 11, 1, 32, 2–10.

47 Gussow, pp. 104–5, 111–3; Jonathan Hutchinson, *On Leprosy and Fish-Eating*, London, 1906, p. 76, quoted in Ibid., p. 106; Gussow and Tracy, p. 436.

48 H.P. Wright, *Leprosy an Imperial Danger*, London, 1889, pp. ix, 4, 16, 18.

49 Brody, pp. 52, 98; Gussow, p. 113.

50 *Papers*, p. 26.

51 *JLIC*, No.1, August 1890, London, pp. 5–6, 9.

52 *Papers*, pp. 295, 294, 20.

53 MacLeod, 'On Visiting the "Moving Metropolis"', p. 2; Harrison, p. 58.

54 Despatch, Sec State, India to GoI (Stats & Commerce), 28 February 1878, Encl to C. E. Buckland, Under-Sec, GoI, H (Med), to CSG, 4 June 1878, P, 42, 14 June 1878; G.O.43, P, 14 June 1878, TNA.

7 Confining leprosy sufferers: the Leprosy Act

1 *Papers*, p. 13.

2 Foucault, *Madness and Civilisation*, p. 39.

3 Arnold, 'Touching the Body', p. 84; *Papers*, p. 86.

4 R. Suntharalingam, *Politics and Nationalist Awakening in South India, 1852–1891*, Tucson, 1974, pp. 190–2; 159–60, 207–10, 352–5, 105, 230.

5 John J. Paul, *The Legal Profession in Colonial South India*, Delhi, 1991, pp. 120–1.

6 *Papers*, pp. 22.

7 Ibid., pp. 11.

8 Harrison, pp. 72–5; Arnold, 'Touching the Body', p. 58.

9 *Papers*, pp. 13, 362.

10 Arnold, 'European Orphans and Vagrants', pp. 104–27; Maclean, *Manual*, Vol. I, pp. 185–6.

11 Leprosy Bill, 1889, Sn 2, Pt 1. (The 1889 Leprosy Bill is available in *Papers*, pp. 14–15.)

12 *Papers*, pp. 37, 40–1, 30.

13 Leprosy Bill, 1889, Sn 2, Pt 2; Sn 3, Pt 2.

14 *Papers*, pp. 11, 31.

15 Leprosy Bill, 1889, Sn 5; Sn 4, Pt 1.
16 *Papers*, p. 21.
17 Ibid., p. 23.
18 Ibid., pp. 36, 26, 35.
19 Ibid., p. 36.
20 *Dinavartamani*, 21 November 1874, in *NNR (Mad)*, November 1874, p. 91, No. 2. I am indebted to Geoff Oddie for this reference.
21 *Papers*, pp. 36, 38–9, 25–6, 34.
22 Ibid., p. 32.
23 Ibid., p. 11.
24 Carter, 'Further Observations', p. 2.
25 *Papers*, pp. 44, 35–36.
26 Day, p. 288.
27 *Madras Times*, 31 January 1884, in Suntharalingam, p. 192; *Papers*, p. 45.
28 Suntharalingam, pp. 352–5.
29 *Papers*, p. 31.
30 Ibid., pp. 3, 28, 32.
31 Ibid., p. 41.
32 Ibid., 23.
33 Arnold, 'Touching the Body', p. 88.
34 Yang, pp. 10–11; Arnold, 'European Orphans and Vagrants', pp. 120–1.
35 *Papers*, p. 24.
36 Suntharalingam, pp. 252–3.
37 *Papers*, pp. 23, 41.
38 Ibid., p. 26.
39 Ibid., pp. 20–1.
40 Ibid., pp. 18–20.
41 Ibid., pp. 299–300.
42 Ibid., p. 299.
43 *JLIC*, No. 1, August 1890, London, p. 8; No. 2, February 1891, pp. 6–7; *Papers*, p. 293.
44 Ibid., pp. 303–4.
45 Ibid.
46 Special Committee members were Geo. N. Curzon, MP, then under-secretary of state for India (chairman), and Edward Clifford, nominated by the Executive Committee of the National Leprosy Fund, Sir Dyce Duckworth and G.A. Heron, nominated by the RCP and Jonathan Hutchinson and N.C. Macnamara, nominated by the Royal College of Surgeons. Ibid., p. 363.
47 Ibid., pp. 304–5.
48 Ibid., p. 364. The other members of the National Leprosy Fund Executive Committee were Baron Ferdinand de Rothschild, MP, the Bishop of London, Sir Algernon Borthwick MP, Sir Edward Lawson, Sir Somers Vine. Ibid., p. 363.
49 'Proposals for giving practical effect in India to the conclusions of the Leprosy Commissioner', 16 January 1895, note, H (Med), 80–117, A, April 1895, NAI.
50 *Papers*, pp. 320, 364–6.
51 Ibid., 364–6.

52 'Lepers Bill', Appendix A, H (Leg), 1–33, A, February 1898; 'Statement of Objects and Reasons', [Lepers Act, 1896], 30 July 1896, H (Med), 34, A, September 1896; J.F. Price, CSG, to Sec, GoI, 10 August 1896, H (Med), 45, A, September 1896, NAI.

53 J.N. Atkinson, Coll, Kistna, to CSG, 10 October 1896, G.O.336M, L & M, 1 March 1897, TNA.

54 F.A. Nicholson, n.d., note, H (Leg), 1–33, A, February 1898, NAI.

55 Lieut-Col, W.A. Lee, Surg, to Sec, Surg-Gen, G.O.336M, L & M, 1 March 1897, TNA.

56 Leprosy Bill, 1896, Sns 9,10. [available at H (Med), 32, September 1896, NAI.]

57 Dist Mag, Madura, to GS, L & M, 5 November 1896; Surg-Gen to GS, L & M, 6 November 1896, G.O.336M, L & M, 1 March 1897; G.S. Forbes, GS, to Sec, GoI, (Leg), 1 March 1897, L & M, 337M, 1 March 1897, TNA. [G.O.336M-337M also in H (Leg), 1–33, A, February 1898, Appendix N1/2, NAI.]

58 Leprosy Bill, 1896, Sn 9, (c),(d).

59 G.W. Elphinstone, Dist Mag, Godavari, to CSG, 16 October 1896, G.O.336M, L & M, 1 March 1897, TNA.

60 G.O.792 Railways, 4 October 1879, quoted in Agent, SIR, to Ctg Eng for Rlwys, 5 February 1890 and 9 January 1890, note in Jdl (Ord), 524, 20 March 1890, TNA.

61 Muhammad Raza Khan, Joint Mag, Sth Arcot, to Dist Mag, Sth Arcot, 3 November 1896, G.O.336M, L & M, 1 March 1897; G.S. Forbes, GS, to Sec, GoI, (Leg), 1 March 1897, L & M, 337M, 1 March 1897, TNA.

62 Surg-Maj W.B. Browning, Surg to Governor, to Sec, Surg-Gen, 26 October 1896, G.O.336M, L & M, 1 March 1897, TNA.

63 Opinion of C. Jambulingam Mudaliyar and J.N. Atkinson, Coll, Kistna, to CSG, 10 October 1896, and N. Subbarao Pantulu Garu to GS, (L & M), 7 November 1896, in G.O.336M, L & M, 1 March 1897, TNA.

64 G.S. Forbes, GS, to Sec, GoI, (Leg), 1 March 1897, L & M, 337M, 1 March 1897, TNA.

65 Lepers Act, 1898, Sn IX, Pt 1, 2; Sn XII.

66 Rogers and Muir, pp. 107–8.

67 S. Swaminatha Aiyar, to CSG, 24 October 1896, G.O.336M, L & M, 1 March 1897, TNA.

68 G.O.835, P, 13 September 1910; P.L. Moore, Pres, Madras Corp, to CSG, 23 December 1910 and G.O. in G.O.89, P, 22 January 1913, TNA.

69 'Detailed statement of prosecutions . . .', P.L. Moore, Pres, Madras Corp, to CSG, 11 July 1911; Col J. Smyth, Surg-Gen, Madras, to CSG, 23 October 1912 and G.O. in G.O.89, P, 22 January 1913, TNA.

70 B. Rama Rao, Under-Sec, to Surg-Gen, Memo, 3682-1 (Med), 8 December 1920, G.O.836, Mad L Self Gov (PH), 16 July 1921, OIOC; R. G. Cochrane, *A Practical Textbook of Leprosy*, London, 1947, pp. 241, 214.

71 S.J. Thomson, SC, UP, 'Note on the present incidence of leprosy in the United Provinces . . .', 28/7; Rev J. Hahn, Spdt, G.S.L Mission, to W. H.P. Anderson, Spdt Chandkhuri, 2 September 1907, 191/7; MTL, 'Annual Report', 1913, p. 10, 25/1, TLM.

72 G.S. Forbes, GS, to Sec, GoI, (Leg), 1 March 1897, L & M, 337M, 1 March 1897, TNA; Rev Frank Oldrieve, 'Legislation Providing for the Care and

Control of Lepers' in MTL, *Report of a Conference . . ., Calcutta, 1920,*
Cuttack, 1920, pp. 60–1.

73 Rev Frank Oldrieve, Sec, Mission to Lepers, to Sec, GoI, (H), 15 October 1919,
    in Ibid., p. 140; Lepers Amended Act, 1920, Sn 2, Pt 1; Cochrane, p. 214.

74 Mr A. Donald Miller, 'Review of the Position Generally of Work among
    Lepers in India', in MTL, *Report of a Conference . . ., Allahabad, 1924,*
    Cuttack, n.d., p. 12.

75 MTL, *Annual Report*, 1913, 25/1, TLM, pp. 9–15.

76 Gussow, p. 208.

77 G.A. Oddie, 'White Rajas: Protestant Missionaries and Imperial Rulers
    in India to 1947', in Deryck M. Schreuder (ed.), *'Imperialisms': Explora-
    tions in European Expansion and Empire*, Sydney, 1991, p. 85.

78 Donald A. Miller, *An Inn Called Welcome: the Story of the Mission to
    Lepers, 1874–1917*, London, 1965, pp. vii–viii.

79 Ignatieff, pp. 4–5.

Biographies

1   Crawford, *Roll*, p. 275; Crawford, *Indian Medical Service*, Vol. II, pp. 22,
    212.

2   Mettler, pp. 142–4; W.F. Bynum and Roy Porter (eds), *Companion Ency-
    clopedia of the History of Medicine*, London 1993, Vol. I, pp. 142–4.

3   D.G. Crawford, *Roll*, p. 468.

4   Ibid., p. 343; Harrison, p. 240.

5   Crawford, *Roll*, p. 138.

6   Crawford, *Indian Medical Service*, Vol. II, p. 416.

7   Until 1818, accounts in Madras were kept in pagodas, fanams and cash.
    80 cash = 1 fanam; 42 fanams = 1 pagoda. In 1818 the rupee became
    the standard coin and the pagoda was reckoned as equivalent to 3.5
    rupees. Maclean, *Glossary*, p. 643.

8   Crawford, *Roll*, p. 352; A.P. Howell, Sec, H (Med), to D.F. Carmichael,
    CSG, 2 August 1875, P, 61, 12 August 1875; E.G. Balfour, Surg-Gen,
    IMD, to D.F. Carmichael, CSG, Madras, 3 September 1875, P, 33, 9 Sep-
    tember 1875, TNA.

9   *Lives of the Fellows of the Royal College of Physicians of London, Munk's
    Roll, 1826–1925*, compiled G.H. Brown, London, 1955, pp. 193–4.

10 Crawford, *Roll*, p. 311.

11 Arnold, *Colonizing the Body*, p. 296, n. 20.

12 *Munk's Roll*, pp. 71–2.

13 Crawford, *Roll*, p. 106.

14 *ICR*, 1901, pp. xvi–ii.

15 Rock, p. 29; Crawford, *Indian Medical Service*, Vol. I, p. 277; Vol. II, pp.
    21, 143–4.

16 Crawford, *Roll*, p. 181; Harrison, p. 253, n. 58.

17 *CR, 1911*, Bengal, Bihar and Orissa, and Sikkim, p. 426.

18 Crawford, *Roll*, p. 339.

Bibliography

ARCHIVES

National Archives of India, New Delhi

Home (Jails)
Home (Judicial)
Home (Medical)
Home (Public)
Home (Sanitary)
Home, Revenue and Agricultural (Medical)
Legislative

Tamil Nadu Archives, Madras

Judicial
Legislative
Local and Municipal
Public
Public Health

Oriental and India Office Collections, London

Board's Collections
Despatches to India
Madras Public
Madras Judicial
Madras Local and Municipal (Medical)
Madras Local Self Government (Public Health)

Wellcome Institute for the History of Medicine

Carter Papers

The Leprosy Mission, London, Archive and Library

Annual Reports
Miscellaneous papers relating to leprosy in India

TAMIL SOURCES

Akattiyar, *Kaṉmakāntam 300*, Palani taṉṭāyutapāṇi cuvāmi tirukkōyil cita maruttuvanūl veḷiyūittuk kuḻu, Madras, 1976.
Akattiyar paḷḷu, Muslim apimaṉi acciyantiracalai, Madras, 1907.
Carapēntira vaitiya muṟaikaḷ, Tanjore, 1965.
Kuppuccāmi Mudaliyār, K.N., *Citta maruttuvam*, Part I, Superintendent Government Printing, Madras, 1954.

Uttamarāyan, K.C., *Cittar aṟuvai maruttuvam*, Tamilnadu Government Printers, Madras, 1968.

Yūkimuṉivar, *Vaittiya cintāmaṇi (Perunūl 800)*, Part I, Tiyākarācan, Irā, Madras, 1976.

Muhammed Abdulla Sahib, *Ūnāni vaittiya thathu virthi bodhini*, Part II, Madras, 1893.

OFFICIAL REPORTS AND PUBLICATIONS

Annual Report of the Sanitary Commissioner for Madras, 1875, Appendix I, 'Leprosy', pp. i–xxii.

Annual Reports of the Civil Dispensaries of the Madras Presidency.

Annual Reports on the Civil Hospitals and Dispensaries of the Madras Presidency.

Carter, H.V., 'Further Observations on the Prevention of Leprosy by Segregation of the affected', Bombay General, 29 June 1887.

Carter, H.V., 'Memorandum on the Prevention of Leprosy by Segregation of the affected', Bombay General, 5 March 1884.

Carter, H.V., 'The Memorial of H.V. Carter, to Kimberley, Sec State, India, 18 August 1884', Carter Papers, *Western Manuscripts*, 5809–5826, 2/1, WIHM.

Carter, H.V., *Report on Leprosy and Leper-Asylums in Norway: With Reference to India*, G.E. Eyre & W. Spottiswoode, London, 1874.

Civil Medical Code Madras, Superintendent of Government Press, Madras, 1898.

Civil Medical Code Madras, 2 Vols, 4th edn, Superintendent of Government Press, Madras, 1929.

Correspondence Relating to the Discovery of an Alleged Cure of Leprosy, G.B., Hse of Commons Papers, 1871, c.362, Vol. LVI, pp. 1–44.

Crawford, F.J., *Syllabus of the Course of Lectures on Materia Medica and Their Therapeutics in the Madras Medical College*, Superintendent of Government Press, Madras, 1895; 1897; 1898.

Hunter, A., 'Report upon the Hydrocotyle Asiatica', *Medical Board Reports Selected by the Medical Board from the Records of their Office*, Christian Knowledge Society Press, Madras, 1855, pp. 356–76.

Leprosy Commission, *Leprosy in India: Report of the Leprosy Commission in India, 1890–91*, Superintendent of Government Printing, Calcutta, 1892.

Lewis, T.R., and D.D. Cunningham, *Leprosy in India*, Calcutta, 1877.

Maclean, C.D. (ed.), *Manual of the Administration of the Madras Presidency*, 3 Vols, Madras, 1885–93 (Asian Educational Services, New Delhi, 1987–90).

Medical Reports Selected by the Medical Board from the Records of their Office, Christian Knowledge Society's Press, Madras, 1855.

Memorandum on the Census of British India of 1871–72, London, 1875.

Milroy, Gavin, *Report on Leprosy and Yaws in the West Indies*, William Clowes & Sons, London, 1873.

Papers Relating to the Treament of Leprosy in India from 1887–1895 (Selections from the Records of the Government of India, Home, No. CCCXXXI), Superintendent of Government Printing, Calcutta, 1896.

Report on the Treatment of Leprosy with Gurjon Oil and Other Remedies in Hospitals of the Madras Presidency (Selections from the Records of the Madras Government, No. Lll.), Government Press, Madras, 1876.

Reports Forwarded by Surgeon-General E.G. Balfour on Certain Forms of Skin Diseases Observed in the Madras Presidency. As suggested in the Pamphlet by T. Fox and T. Farquar etc. (Supplementary Report), Office of the Surgeon-General, Madras, 1875.

Reports on Medical Topography and Statistics, compiled from the Records of the Medical Board Office, Madras.

Royal College of Physicians, *Report on Leprosy*, George Edward Eyre & William Spottiswoode, London, 1867.

Standing Orders of the Government Leper Hospital Madras, 1895, Superintendent Government Press, Madras, 1896.

Standing Orders of the Government Ophthalmic Hospital, Madras, 1895, Superintendent Government Press, Madras, 1895.

SERIALS

Census of India
Fort St George Gazette
Indian Annals of Medical Science
Indian Medical Gazette
Indian Medical Record
Journal of the Leprosy Investigation Committee
Journal of Tropical Medicine
Lancet
Leprosy in India
Madras Journal of Medical Science
Madras Medical Journal
Madras Quarterly Journal of Medical Science
Madras Quarterly Medical Journal
Madras Times
Medical Missions in India
Native Newspaper Reports (Madras)
Transactions of the Medical and Physical Society of Bombay

PRIMARY SOURCES

Ainslie, Whitelaw, *Materia Indica*, 2 Vols, London, 1826 (Vol. I, Periodical Expert Book Agency, Delhi, 1986); (Vol. II, Neeraj Publishing House, Delhi, 1984).

Bailley, Wellesley C., *A Glimpse at the Indian Mission-Field and Leper Asylums in 1886–7*, John F. Shaw & Co., London, 1890 edn.

Baillie, Neil B. E., *A Digest of Moohummudan Law, on the subjects to which it is usually applied by British Courts of Justice in India, compiled and translated from Authorities in the Original Arabic*, Part II, Smith Elder & Co., London, 1869.

Birch, Edward A., 'Gleanings from a Mofussil Practice', *Indian Medical Gazette*, 1 July 1878, pp. 180–3.

Bruce, Mitchell J., *Materia Medica and Therapeutics*, 3rd edn, Cassell, London, 1886.

Carter, H.V., 'On the Symptoms and Morbid Anatomy of Leprosy; with Remarks', *Transactions of the Medical and Physical Society of Bombay*, No. VIII, New Series, 1862 (Bombay, 1863), pp. 1–105.

Day, Francis, 'Elephantiasis Grœcorum, or Leprosy', *Madras Quarterly Journal of Medical Science*, I (1860) pp. 286–300.

Dougall, J., *Report on the Treatment of Leprosy with Gurjon Oil, with Other Papers; Extracted from the 'Indian Medical Gazette'*, Calcutta, 1876. [Tract, 516, OIOC]

Fleming, J.M., 'On the Curability of Leprosy', *Indian Medical Gazette*, 2 January 1871, pp. 14–15; 1 March 1871, pp. 53–4; 1 June 1871, pp. 119–20.

Fox, Tilbury, and T. Farquhar, 'Abstract of the Original Scheme for Obtaining a Better Knowledge of the Endemic Skin Diseases of India', in Fox, Tilbury, and T. Farquhar, *On Certain Endemic Skin and Other Diseases...*, Appendix I, pp. 1–31.

Fox, Tilbury, and T. Farquhar, *On Certain Endemic Skin and Other Diseases of India and Hot Climates Generally*, J. & A. Churchill, London, 1876.

Ghose, J.C., *A Monograph on Chaulmoogra Oil and its Use in the Treatment of Leprosy as Explained in Ayurveda*, pub. by the author, Madras, 1917.

Gupta, K.K., 'An Easy Method of Keeping Hydnocarpus Oil Constantly Hot During Injection', *Leprosy in India*, July 1934, p. 136.

Jee, Bhagavat Singh, *A Short History of Aryan Medical Science*, Macmillan & Co., London, 1896.

Johnston, W., 'Observations on Leprosy and on its Treatment by Means of Vaporized Cabolic Acid in Union with Watery Vapour', *Indian Medical Gazette*, 2 November 1874, pp. 287–9; 1 January 1875, pp. 9–12.

Lawrie, Edward, 'Nerve Stretching in Anæsthetic Leprosy', *Indian Medical Gazette*, 2 September 1878, p. 229; 1 October 1878, p. 270.

Love, Henry Davison, *Vestiges of Old Madras, 1640–1800*, n.d. 4 Vols, (Mittal Publications, Delhi, 1988).

Macnaghten, Sir William Hay, *Principles and Precedents of Moohummudan Law*, 2nd edn, (with additional notes by William Sloan), Higginbotham & Co., Madras, 1860.

Mayne, John D., *A Treatise on Hindu Law and Usage*, 9th edn, Higginbotham & Co., Madras, 1922.

Mouat, F.J., 'Notes on Native Remedies', *Indian Annals of Medical Science*, October 1853 and April 1854, I (1854) pp. 646–52.

Nelson, J.H., *A View of the Hindu Law: as Administered by the High Court of Judicature at Madras*, Higginbotham & Co., Madras, 1877.

Neve, Arthur, 'Leprosy and Christianity', *Journal of Tropical Medicine*, 15 May, VIII (1905) pp. 146–7.

Playfair, George, *The Taleef Shereef, or Indian Materia Medica*, Baptist Mission Press, Calcutta, 1883 (Bishen Singh Mahendra Pal Singh, Dehra Dun, 1984).

Roy, G.C., 'Some Remarks on Leprosy', *Indian Medical Gazette*, 1 February 1881, pp. 45–6.

Royle, J. Forbes, *A Manual of Materia Medica and Therapeutics*, 3rd edn, (rev. F.W. Headland), John Churchill, London, 1856.

Sastri, Krishna H., *South-Indian Images of Gods and Goddesses*, Madras, 1916 (Asian Educational Services, New Delhi, 1986).

'The Edinburgh Medical Journal on "Sanitary Science" and Medical "Statistics"', *Madras Quarterly Journal of Medical Science*, II (1861) pp. 197–9.

'The History of Leprosy', *Madras Quarterly Journal of Medical Science*, II (1861) pp. 192–5.

The Ordinances of Manu, trans. Arthur Coke Burnell, ed. Edward W. Hopkins, London, 1884 (Oriental Books Reprint Corporation, 2nd edn, New Delhi, 1971).

Van-Someren, W.J., 'A Brief Historical Sketch of the Madras Leper Hospital, with some Notices of Leprosy as it is seen in that Institution', *Madras Quarterly Journal of Medical Science*, III (1861) pp. 271–94.

Wallace, James R., 'Nerve-Stretching in Anæsthetic Leprosy', *Indian Medical Gazette*, 1 November 1880, pp. 300–1.

Waring, Edward John, *Pharmacopœia of India*, India Office, London, 1868.

Whitehead, Henry, *The Village Gods of South India*, 2nd edn, Madras, 1921 (Asian Educational Services, New Delhi, 1988).

Williams, Monier, *Religious Thought and Life in India: Vedism, Brāhmanism and Hindūism*, London, 1883 (Oriental Books Reprint Corporation, New Delhi, 1974).

Wilson, H.H., 'Kushta, or Leprosy: as Known to the Hindus', *Transactions of the Medical and Physical Society of Calcutta*, 1 and 3 May 1823, pp. 1–44.

Wright, H.P., *Leprosy an Imperial Danger*, J. & A. Churchill, London, 1889.

SECONDARY SOURCES

Andiappa Pillai, D., 'Kalluppu in Muppu' [sic], in Subramanian and Madhavan (eds), pp. 146–53.

Arnold, David, *Colonizing the Body: State Medicine and Epidemic Disease in Nineteenth-Century India*, University of California Press, Berkeley, 1993.

Arnold, David, 'Crisis and Contradiction in India's Public Health', in Dorothy Porter (ed.), *The History of Public Health and the Modern State*, Rodopi, Amsterdam, Atlanta, 1994, pp. 335–55.

Arnold, David, 'European Orphans and Vagrants in India in the Nineteenth Century', *Journal of Imperial and Commonwealth History*, January, VII, 2 (1979) pp. 104–27.

Arnold, David (ed.), *Imperial Medicine and Indigenous Societies*, Oxford University Press, Delhi, 1989.

Arnold, David, 'Medical Priorities and Practice in Nineteenth-Century British India', *South Asia Research*, 5, 2 (1985) pp. 167–83.

Arnold, David, 'Smallpox and Colonial Medicine in Nineteenth-Century India', in Arnold (ed.), pp. 45–65.

Arnold, David, 'Touching the Body: Perspectives on the Indian Plague, 1896–1900', *Subaltern Studies V*, Oxford University Press, Delhi, 1990, pp. 55–90.

Bala, Poonam, *Imperialism and Medicine in Bengal: a Socio-Historical Perspective*, Sage, New Delhi, 1991.

Ball, James Moores, *The Body Snatchers*, Dorset Press, New York, 1989.

Barrier, N. Gerald (ed.), *The Census in British India: New Perspectives*, Manohar, New Delhi, 1981.

Basalla, George, 'The Spread of Western Science', *Science*, 156 (1967) pp. 611–22.

Bhatia, S.L., *A History of Medicine*, Medical Council of India, New Delhi, 1977.

Bose, D.M., S.N. Sen and B.V. Subbarayappa (eds), *A Concise History of Science in India*, Indian National Science Academy, New Delhi, 1971.

Brody, Saul, Nathaniel, *The Disease of the Soul: Leprosy in Medieval Literature*, Cornell University Press, Ithaca, 1974.

Buckingham, Jane, 'The "Morbid Mark": the Place of the Leprosy Sufferer in Nineteenth Century Hindu Law', *South Asia*, XX, 1 (1997) pp. 57–80.

Bynum, W.F., *Science and the Practice of Medicine in the Nineteenth Century*, Cambridge University Press, Cambridge, 1994.

Bynum, W.F. and Roy Porter (eds), *Companion Encyclopaedia of the History of Medicine*, 2 Vols, Routledge, London, 1994.

Cartwright, F.F., 'Antiseptic Surgery', in Poynter (ed.), pp. 77–103.

Christian, M., 'Epidemiology', in Thangaraj (ed.), *A Manual of Leprosy*, pp. 4–30.

Cochrane, R.G., *A Practical Textbook of Leprosy*, Oxford University Press, London, 1947.

Cooke, A.M., *A History of the Royal College of Physicians of London*, Vol. III, Clarendon Press, Oxford, 1972.

Crawford, D.G., *A History of the Indian Medical Service, 1600–1913*, 2 Vols, Thacker & Co., London, 1914.

Crawford, D.G., *Roll of the Indian Medical Service, 1615–1930*, Thacker & Co., London, 1930.

Da Silva Gracias, Fátima, *Health and Hygiene in Colonial Goa, 1510–1961*, XCHR Studies Series No. 4, Concept Publishing Company, New Delhi, 1994.

Daniel, E. Valentine, 'The Pulse as an Icon in Siddha Medicine', in E. Valentine Daniel and Judy F. Pugh (eds), *South Asian Systems of Healing, Contributions to Asian Studies*, XVIII (1984) pp. 115–26.

Davey, Cyril, *Caring Comes First: the Leprosy Mission Story*, Marshall Pickering, Basingstoke, 1987.

Derrett, J.D.M., 'J.H. Nelson: a Forgotten Administrator-Historian of India', in C.H. Philips (ed), *Historians of India, Pakistan and Ceylon*, Oxford University Press, London, 1961, pp. 354–72.

Derrett, J.D.M., *Religion, Law and the State in India*, Faber & Faber, London, 1968.

Ernst, Waltraud, 'The Establishment of "Native Lunatic Asylums" in Early Nineteenth-Century British India', in Meulenbeld and Wujastyk (eds), pp. 169–202.

Foucault, Michel, *Discipline and Punish: the Birth of the Prison*, Peregrine, 1979.

Foucault, Michel, *Madness and Civilisation: a History of Insanity in the Age of Reason*, Vintage, New York, 1988.

Gaitonde, P.D., *Portuguese Pioneers in India: Spotlight on Medicine*, Popular Prakashan, Bombay, 1983.

Ganapati, R., 'Therapy of Leprosy', in Thangaraj (ed.), pp. 143–55.

Geetha, G., 'Kaya Kalpa Mooligai in Siddha Medicine', in Subramanian and Madhavan (eds), pp. 132–45.

Goffman, Erving, *Asylums: Essays on the Social Situation of Mental Patients and Other Inmates*, Peregrine, 1987.

Gussow, Zachary, *Leprosy, Racism, and Public Health: Social Policy in Chronic Disease Control*, Westview Press, London, 1989.

Gussow, Zachary, and George S. Tracy, 'Stigma and the Leprosy Phenomenon: the Social History of a Disease in the Nineteenth and Twentieth Centuries', *Bulletin of the History of Medicine*, XLIV (1970) pp. 425–49.

Harrison, Mark, *Public Health in British India: Anglo-Indian Preventive Medicine, 1859–1914*, Cambridge University Press, Cambridge, 1994.

Hastings, Robert C. (ed.), *Leprosy*, 2nd edn, Churchill Livingston, London, 1994.

Headrick, Daniel R., *The Tools of Empire: Technology and European Imperialism in the Nineteenth Century*, Oxford University Press, Oxford, 1981.

Hodgkinson, Ruth G., 'Social Medicine and the Growth of Statistical Information', in Poynter (ed.), pp. 183–98.

Hume, John Chandler, 'Colonialism and Sanitary Medicine: the Development of Preventive Health Policy in the Punjab, 1860–1900', *Modern Asian Studies*, 20, 4 (1986) pp. 703–24.

Ignatieff, M., *A Just Measure of Pain: the Penitentiary in the Industrial Revolution*, Macmillan, London, 1978.

Iliffe, John, *The African Poor: a History*, Cambridge University Press, Cambridge, 1987.

Jaggi, O.P., *Indian System of Medicine*, History of Science and Technology in India, Vol. IV, Atma Ram, Delhi, 1973.

Jois, Rama M., *Legal and Constitutional History of India*, 2 Vols, N.M. Tripathi, Bombay, 1984.

Jolly, Julius, *Indian Medicine*, 2nd edn, Munishiram Manoharlal, New Delhi, 1977.

Kandaswamy Pillai, N., *History of Siddha Medicine*, Government of Tamil Nadu, Madras, 1979.

Krishnan, K.R., 'Siddha Medicine During the Period of the Marattias' [sic], in Subramanian and Madhavan (eds), pp. 54–86.

Kumar, Deepak (ed.), *Science and Empire: Essays in Indian Context (1700–1947)*, Anamika Prakashan, New Delhi, 1991.

Leslie, Charles, 'The Ambiguities of Medical Revivalism in Modern India', in C. Leslie (ed.), *Asian Medical Systems: a Comparative Study*, University of California Press, Los Angeles, 1976, pp. 356–67.

MacKenzie, Donald A., *Statistics in Britain, 1865–1930: the Social Construction of Scientific Knowledge*, Edinburgh University Press, Edinburgh, 1981.

MacLeod, Roy, 'On Visiting the "Moving Metropolis"; Reflections on the Architecture of Imperial Science', *Historical Records of Australian Science*, 5, 3 (1982) pp. 1–16.

Manickavasagam, R., 'Contribution of Agathiyar to Siddha System of Medicine', in Subramanian and Madhavan (eds), pp. 577–611.

Marglin, Frédérique, Apffel, 'Smallpox in Two Systems of Knowledge', in Frédérique Apffel and Stephen A. Marglin (eds), *Dominating Knowledge: Development, Culture, and Resistance*, Clarendon Press, Oxford, 1990, pp. 102–44.

Mettler, Cecilia, C., *History of Medicine*, Blakiston, Philadelphia, 1947.

McGrew, Roderick E., *Encyclopedia of Medical History*, McGraw Hill Book Company, New York, 1985.

Meulenbeld, G. Jan and Wujastyk, Dominik (eds), *Studies on Indian Medical History*, E. Forsten, Groningen, 1987.

Miller, Donald, A., *An Inn Called Welcome: the Story of the Mission to Lepers, 1874–1917*, the Mission to Lepers, London, 1965.

Munk's Roll, Lives of the Fellows of the Royal College of Physicians of London, 1826–1925, (compiled G.H. Brown), Royal College of Physicians, London, 1955.

Narayanaswami, V., 'Ayurveda and Siddha Systems of Medicine – a Comparative Study', in Subramanian and Madhavan (eds), pp. 568–76.

Oddie, Geoffrey, A., '"Orientalism" and British Protestant Missionary Constructions of India in the Nineteenth Century', *South Asia*, XVII, 2 (1994) pp. 27–42.

Oddie, Geoffrey, A., 'White Rajas: Protestant Missionaries and Imperial Rulers in India to 1947', in Deryck M. Schreuder (ed.), *'Imperialisms': Explorations in European Expansion and Empire*, Dept of History, University of Sydney, Sydney, 1991.

Patterson, T.J.S., 'The Relationship of Indian and European Practitioners of Medicine from the Sixteenth Century', in Meulenbeld and Wujastyk (eds), pp. 119–29.

Paul, John J., *The Legal Profession in Colonial South India*, Oxford University Press, Delhi, 1991.

Pearson, M.N., 'The Thin End of the Wedge: Medical Relativities as a Paradigm of Early Modern Indian-European Relations', *Modern Asian Studies*, 29, 1 (1995) pp. 141–70.

Pfaltzgraff, Roy E., and Gopal Ramu, 'Clinical Leprosy', in Hastings (ed.), pp. 237–87.

Poynter, F.N.L. (ed.), *Medicine and Science in the 1860s: Proceedings of the Sixth British Congress on the History of Medicine*, University of Sussex, 6–9 September 1967, Wellcome Institute for the History of Medicine, London, 1968.

Radhakrishnan, K., 'Siddha Medicine for the Skin Diseases' [sic), in Subramanian and Madhavan (ed.), pp. 407–25.

Ramasubban, Radhika, 'Imperial Health in British India, 1857–1900', in Roy MacLeod and Milton Lewis (eds), *Disease, Medicine and Empire, Perspectives on Western Medicine and the Experience of European Expansion*, Routledge, London, 1988, pp. 38–60.

Ramasubban, Radhika, *Public Health and Medical Research in India: Their Origins Under the Impact of British Policy*, SAREC, Stockholm, 1982.

Richardson, Ruth, *Death, Dissection and the Destitute*, Penguin, 1988.

Rock, Joseph F., 'The Chaulmoogra Tree and Some Related Species: a Survey Conducted in Siam, Burma, Assam, and Bengal', US Dept of Agriculture, Bulletin No. 1057, Washington, DC, Professional Paper, 24 April 1922.

Rogers, Leonard, and Ernest Muir, *Leprosy*, 2nd edn, Simkin Marshall, London, 1940.

Rogers, Leonard, 'The Successful Treatment of Leprosy', the *Practitioner*, August, CVII, 2 (1921) pp. 77–101.

Sampath, C.K., 'Evolution and Development of Siddha Medicine', in Subramanian and Madhavan (eds), pp. 1–20.

Smith, F.B., *The People's Health, 1830–1910*, Weidenfeld & Nicolson, London, 1990.

Subbarayappa, B.V., 'Chemical Practices and Alchemy', in Bose, D.M., Sen, S.N. and Subbarayappa, B.V. (eds), *A Concise History of Science in India*, Indian National Science Academy, New Delhi, 1971, pp. 274–349.

Subramanian, S.V., and V.R. Madhavan (eds), *Heritage of the Tamils: Siddha Medicine*, International Institute of Tamil Studies, Madras, 1983.

Suntharalingam, R., *Politics and Nationalist Awakening in South India, 1852–1891*, University of Arizona Press, Arizona, 1974.

Suresh, A., and G. Veluchamy, 'Surgery in Siddha System' [sic], in Subramanian and Madhavan (eds), pp. 461–71.

Tamil Lexicon, University of Madras, Madras, 6 Vols and Supplement, 1982.

Thangaraj, R.J. (ed.), *A Manual of Leprosy*, 6th edn, Printaid, New Delhi, 1989.

Trautman, John R., 'The History of Leprosy', in Hastings (ed), *Leprosy*, pp. 11–25.

Vaughan, Megan, *Curing Their Ills: Colonial Power and African Illness*, Polity Press, Cambridge, 1991.

Virchow, Rudolf, *Cellular Pathology: as based upon Physiological and Pathological Histology*, 2nd German edn, Frank Chance (trans), Dover, N.Y., 1971.

Yang, Anand A. (ed.), *Crime and Criminality in British India*, University of Arizona Press, Tucson, 1985.

Yule, Henry and A.C. Burnell, *Hobson-Jobson, A Glossary of Colloquial Anglo-Indian Words and Phrases, and of Kindred Terms, Etymological, Historical, Geographical and Discursive*, William Crooke (ed.), 2nd ed., Munshiram Manoharlal, Delhi, 1968.

Zimmermann, Francis, *The Jungle and the Aroma of Meats: an Ecological Theme in Hindu Medicine*, University of California Press, London, 1987.

Zvelebil, Kamil V., *Companion Studies to the History of Tamil Literature*, E.J. Brill, Leiden, 1992.

Zvelebil, Kamil V., *The Poets of the Powers*, Rider & Co., London, 1973.

Zvelebil, Kamil V., *Tamil Literature*, Harrowitz, Wiesbaden, 1974.

Zvelebil, Kamil, *The Smile of Murugan on Tamil Literature of Southern India*, E.J. Brill, Leiden, 1973.

Index